The Evolution
of Public Policy

THE
EVOLUTION
OF PUBLIC
POLICY

Cars and the Environment

TONI MARZOTTO
VICKY MOSHIER BURNOR
GORDON SCOTT BONHAM

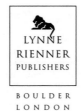

LYNNE
RIENNER
PUBLISHERS

BOULDER
LONDON

Published in the United States of America in 2000 by
Lynne Rienner Publishers, Inc.
1800 30th Street, Boulder, Colorado 80301
www.rienner.com

and in the United Kingdom by
Lynne Rienner Publishers, Inc.
3 Henrietta Street, Covent Garden, London WC2E 8LU

Library of Congress Cataloging-in-Publication Data
Marzotto, Toni, 1945–
    The evolution of public policy :   cars and the environment   /   by
Toni Marzotto, Vicky Moshier Burnor, Gordon Scott Bonham.
        p.   cm.
    Includes bibliographical references and index.
    ISBN 1-55587-858-X (hc. :   alk. paper)
    ISBN 1-55587-882-2 (pbk. :   alk. paper)
        1. Transportation, Automotive—Environmental aspects—United
States.   2. Commuting—Government policy—United States.
3. Environmental policy—United States—Decision making.
I. Burnor, Vicky Moshier, 1955–   . II. Bonham, Gordon Scott.
III. Title.
TD195.T7M37   2000
363.739'256'0973—dc21   99-25923
                                                                                        CIP
                                                                                        Rev.

British Cataloguing in Publication Data
A Cataloguing in Publication record for this book
is available from the British Library.

Printed and bound in the United States of America

         The paper used in this publication meets the requirements
  ∞   of the American National Standard for Permanence of
         Paper for Printed Library Materials Z39.48-1984.

    5 4 3 2 1

For Bill, Christopher, and Carla Jackson;
Rick, Christopher, John, Jeffrey, and Elisabeth Burnor;
and Sandra Bonham

Contents

Illustrations

Tables

Figures

Boxes

Preface

Like it or not, our beloved car is an irksome source of pollution, urban congestion, and excessive fossil-fuel consumption. I am no Greenpeacer, but . . .

—Brock Yates (1988)

The above statement, written in 1988 by a columnist for *Car and Driver,* a magazine popular among motorists, reflected the growing recognition among many segments of society that the American love affair with the automobile needed reexamining. A year later, an editorial in *Sport Truck* conceded that "even if performance has to be compromised by clean air legislation, we are just going to have to bite the bullet. Because when our enthusiasm butts heads with our health, something has to give" (1989, 18). It was in this atmosphere of renewed interest in the linkage between cars and the environment that the 1990 Clean Air Act Amendments (CAAA-90) emerged. Although this linkage is still heavily informed by the belief that technology can produce cars that are friendlier to the environment, there is growing realization that behavioral changes will also be needed. In 1990, Congress agreed with environmentalists and defined solo commuting as environmentally harmful behavior. The CAAA-90 mandated that large companies in metropolitan areas with severe ozone pollution should reduce the number of their employees who drive alone to work. This mandate, the employee commute options (ECO), seemed simple and logical but soon proved to be anything but. Asking people to leave their beloved cars at home was tantamount to heresy, but the Clean Air Act did just that.

In this book, we attempt to analyze the policy process and understand what happened when government tried to get citizens to stop driving so much. We try to answer the question, "Why do good ideas not always make good policy?"

Objectives

We had three objectives in writing this book. Our first objective was to analyze a specific policy as it moves from idea through implementation and finally to impact. Most books in the policy analysis field focus on either policy formulation or policy implementation. Although both are important, when looked at separately neither gives students the complete picture of the policy process. Since ECO moved quickly through the policymaking process, studying it allows students of public policy to follow one policy all the way through the process. Our book is a rich source of primary material for students and scholars, incorporating many original documents and data—statutes, federal guidance, state regulations, company plans, and employee surveys. We examine the actions of public actors from federal to local levels of government as well as the actions of private companies and individual employees.

Our second objective was to explain the policy process. Our research, funded by the U.S. Environmental Protection Agency, focused on the social and economic impact of ECO on private companies and their employees. We assumed that once promulgated, laws and regulations would remain relatively fixed. However, as we went on, we found that neither laws nor regulations are fixed. Public policies are subject to constant pressure for change. We found that change is an integral part of the policy process.

Our third objective was to analyze why policies change and why some groups are active during the formulation of policy but inactive during its implementation. To help students understand what happens, we integrated and adapted two models—the policy cycle model and the advocacy coalition framework—into a new model, the policy cycle advocacy system. By combining these two models into the policy cycle advocacy system, we believed we made a significant contribution to the literature of policy analysis.

Plan of the Book

Chapter 1 discusses public policy and the policymaking process. It outlines an integrated model that views the process as a cycle with six stages. The integrated model specifies what factors move policies from one stage to the next. The rest of the book applies this model to an analysis of ECO and seeks to extrapolate general principles that are applicable to any policy. The historical context in which issues emerged must be reviewed to understand why they are perceived as problems and begin to move through the policy cycle.

Chapter 2 focuses on how problems are initially perceived and defined. The historical context of air quality issues helped shape the future development of ECO policy as it intersected three different policy subsystems. The control and allocation of common goods lead to conflicts among individuals and groups who have differing basic, policy, and instrument beliefs. Chapter 3 looks at the advocacy coalitions that develop as individuals and groups seek to promote their beliefs. They vie to get their definition of the problem on the government's institutional agenda for the formulation of policy. Chapter 4 focuses on policy formulation. A review of the legislative history of ECO provides the context for understanding the roles played by the two major advocacy coalitions in the three policy subsystems.

The next two chapters discuss policy implementation principles and apply them to ECO. They show how policy goals do not always translate into changes that will solve, or even affect, the problem. Chapter 5 focuses on implementation by government agencies at the federal and state levels. Chapter 6 extends the implementation picture to private companies and individual citizens, both targets of ECO. Our empirical research, using employer and employee surveys, reveals how ECO was, and was not, implemented.

Chapter 7 discusses the role and importance of formal and informal evaluations in the policy cycle and our efforts to evaluate ECO formally. Unfortunately, most public policies take longer to implement fully than policymakers, advocacy coalitions, or the public are willing to wait. The rush to judge policy either a success or failure is hard to avoid but may adversely affect a policy before it has had a chance to work. Public policies rarely end because they have eliminated problems. Neither are they static; they keep changing. Chapter 8 looks at the reformulation of ECO policy from a mandatory to a voluntary program. The withdrawal of government mandates from ECO might have meant that it would drop off the institutional agenda. However, as long as people perceive and define problems that need government attention, advocacy coalitions will continue to drive policies through the various stages of the policy cycle as they gain resources from, and are constrained by, the larger cultural, social, economic, and political systems. Congestion and air pollution are still problems in cities. Where and when will we next see new public policy linking driving behavior to clean air?

* * *

We thank many people who provided information and assistance along the way. We received substantial information from federal and regional officials at the U.S. Environmental Protection Agency, the Delaware Department of Transportation, the Maryland Department of the Environment, the New Jersey Department of Transportation, and the Pennsylvania Department of Natural Resources. The Delaware Valley Regional Planning Council, the Baltimore Metropolitan Council, and transportation management associations in Delaware, Maryland, and New Jersey also provided information. We add a special note of thanks to Connie Ruth, environmental protection specialist at the EPA's Office of Mobile Sources in Ann Arbor, Michigan. Connie helped us understand many of the ECO details, she answered our regular requests for information and updates, and she read and critiqued our manuscript.

Many other people participated in the research and preparation of this book. Andrew Farkas, director of transportation studies at Morgan State University, contributed to the understanding of regional transportation patterns and issues. M. Tom Basuray and Trudy L. Somers in the Department of Management at Towson University brought expertise in business and human resource management. Nancy Horst, Mimi Shi, Linda Heil, and Lori Kessler Burrell provided support as graduate research assistants. Deborah Epps contributed general administrative and secretarial support. Kevin R. Dungey, director, Quality Communications Group, gave valuable editorial comments on chapter drafts. We also thank Emmett N. Lombard, associate professor of political science at Oakland University in Rochester, Michigan, who reviewed the draft of the manuscript and offered valuable suggestions on condensing and focusing the material. We thank the Department of Political Science and the Office of Faculty Development at Towson University for granting Toni Marzotto a full-year sabbatical to complete the manuscript.

The research was supported by a grant from the U.S. Environmental Protection Agency (R821240-01-0) to Towson State University. The "Economic and Social Impact of Employee Commute Requirements of the Federal Clean Air Act Amendments of 1990" project, with Gordon Scott Bonham as principal investigator, began February 1994 and ended January 1998.

Finally, we thank our families for their support during the long period from gestation to completion of the project. Bill, Christopher, and Carla Jackson supported and endured with Toni Marzotto. Rick, Christopher, John, Jeffrey, and Elisabeth were long-suffering with Vicky Burnor's involvement, which spanned two interstate moves. Sandra Bonham accepted completion of this book as a part of Gordon Bonham's transition from university to private business.

—*The Authors*

1

Analyzing Public Policy:
Policy Stages and Advocacy Coalitions

Public policy in the United States affects each citizen in hundreds of ways—some familiar, some unsuspected. Citizens directly confront public policy when they are arrested for speeding, but they seldom remember that the air they breathe or the food they eat is regulated by government agencies. Not so long ago, most citizens routinely tossed aluminum cans in the trash. Today, many communities have laws prohibiting such behavior; some even issue citations or fines to enforce these laws.

Public policy is rarely a single event or decision. Some public policies are clear; some are not. Some policies must be inferred from a series of actions and behaviors of many government agencies, levels of government, and officials involved in policymaking over time. Other policies can be found in the statute books.

Defining Public Policy

Despite these complexities, definitions of public policy abound. One definition holds that "public policy is whatever the government chooses to do or not to do" (Dye 1984, 1). Another definition of policy suggests that it is "a long series of more-or-less related activities" rather than a single discrete decision (Rose 1969, 6). This definition embodies the useful notion that policy is a course or pattern of activity rather than a single decision to do something. We subscribe to a variation on James E. Anderson's (1997) definition: Public policy is an intentional course of action followed by government institutions or officials to resolve an issue of public concern. Such a course of action must be manifested in laws, public statements, official regulations, and publicly visible patterns of government behavior. This definition links policy to goal-oriented activity directed toward resolving a problem rather than to random behavior or chance occurrences. Further, this definition focuses on what governments actually do, not just on what they say they are going to do.

There is no single process by which policies are made. Policies do not come off an assembly line like automobiles or television sets (Anderson 1997, 38). Rather, the variations in the subjects of policies produce variations in the style and techniques of policymaking. Environmental policy, professional licensing, and reform of local government are each characterized by a distinguishable policy process—different participants, procedures, techniques, decision rules, and the like. Policymaking may also vary depending upon whether its primary organizational location is in the legislature, the executive, the judiciary, or an administrative agency (Peters 1996). Common processes or elements, however, permeate the variability between different types of policy.

Employee Commute Options Policy

Employee commute options (ECO) is an interesting study in public policy for both procedural and substantive reasons. From a procedural point of view, ECO moved quickly through the policymaking process from the time it emerged on the national institutional agenda in 1989 through policy formulation and implementation. By 1995 it had been reformulated from a mandatory to a voluntary program, which extended its life. Many jurisdictions have implemented voluntary programs, including most of the areas that previously had mandatory programs. Several conferences have been held on the use of voluntary and episodic programs, and the Environmental Protection Agency (EPA) has issued guidelines on how to set up voluntary commute reduction programs (USEPA 1996) as well as on how states can get credit for voluntary programs in their state clean air implementation plans (USEPA 1997).

Substantively, ECO is an important policy to analyze because it employs a new approach for solving a recurring problem in our society. Rather than rely on technology-forcing strategies, Congress sought to change behavior. It assumed that a reduction in the number of vehicle miles traveled between home and work would improve air quality, reduce congestion, and renew urban space. The problems ECO was formulated to solve have not been solved, nor have they gone away. In fact, these problems are likely to become bigger in the future. ECO was an effort to deal with the intersection of these problems with a single solution. The solution, however, grew in complexity as it moved from theory (problem definition) to practice (implementation).

The intergovernmental features of U.S. federalism are also highlighted in the ECO story. Much of the implementation of national policy takes place at the state and local levels. These governments are the real work-horses of the federal system, providing services as well as enforcing regulations. An examination of the regulations developed by selected states to translate federal ECO policy into the nuts and bolts of the operation reveals the variability and flexibility of U.S. federalism. Each state promulgated different rules in an effort to respond to their own constituents and to maximize their economic competitiveness.

The free and equal access to common, or public, goods like air and water often leads to tension between the desires of the individual and the needs of the community. Public policy is an effort to resolve this tension, but solving a problem in one part of the commons does not solve "the tragedy of the commons" (Coughlin 1994). ECO shows some of the limitations of U.S. federalism when the commons is much larger than jurisdictional boundaries.

The Purpose of the Study

Our examination of the ECO policy in this book provides students of the policymaking process with an opportunity to look at a substantive issue from beginning to end. ECO required that employers of 100 or more employees, in ten metropolitan areas with severe or extreme ozone or carbon monoxide nonattainment, implement plans to increase the average vehicle occupancy during home to work commutes, that is, reduce solo commuting. (See Figure 1.1.) The logic behind ECO seemed self-evident: if employees carpooled or took public transportation, fewer cars on the roads would translate into less pollution, and less pollution would reduce the number of urban areas in noncompliance with federal clean air standards.

The prospect of requiring employers to change the commuting behavior of their employees was fascinating to us. Not only did this provision of the Clean Air Act Amendments of 1990 (CAAA-90) move from previous efforts to force changes in technology (produce cleaner cars) to forcing changes in behavior (make individuals drive less), but it also moved from direct regulation to indirect regulation. Employers were held responsible for the behavior of their employees outside the workplace but were given no new tools to achieve this desired behavioral change. We were particularly interested in how companies could change their employees' behavior, especially a behavior as deeply ingrained as driving to work.

We wanted to study the economic and social impact of policy implementation at the company and employee level, using the tools of business management. Our interest was shared by the USEPA, which gave us a

Figure 1.1   Regions Subject to ECO Regulations

Region	State
New York	Connecticut
	New York
	New Jersey
Philadelphia	Pennsylvania
	Delaware
Baltimore	Maryland
Chicago	Illinois
	Indiana
Milwaukee	Wisconsin
Houston	Texas
Los Angeles	
Ventura County	
Southeastern Desert	California
San Diego	

three-year grant to study the implementation of this policy, first in the Baltimore Metropolitan Region and later in the Philadelphia Metropolitan Region. We began our research with the following assumptions:

• The law would remain unchanged;
• States would comply with federal statutes;
• Employers would comply with state regulations; and
• Employees would change their commuting behavior.

Our focus was on the employers, and we assumed that federal and state implementation of the law would be stable. We soon discovered that the simple passage of federal law does not define policy. States, especially in the 1990s, are not afraid to challenge federal laws. Neither are companies, who frequently take their cues from state governments. Companies comply with laws when they believe there will be consequences if they do not or when they believe it is in their best business interest to do so. We learned that in the world of policy analysis, stability is a rare commodity. Even while our study was getting under way, the situation began to change. Negative media coverage, protests by business leaders, and state challenges eventually led to changes in the law.

For us ECO was a learning experience. We learned that good ideas or good laws don't always translate into good policies and that understanding the policy process requires two sets of lenses—a wide angle lens to provide a big picture of what is going on and a telephoto lens to help zoom in on key events at important moments. We developed a theoretical model to enable us to simulate that metaphorical camera. In our effort to explain what was happening with ECO, we initially turned to the policy cycle model (Anderson 1997), a staple of the policy analysis literature. We found that this model is useful as an organizing tool, especially at the beginning of an analytical endeavor, but it does not have zoom capability. The advocacy coalition framework (Sabatier 1988; Sabatier and Jenkins-Smith 1993), however, focuses intensely on particular policy subsystems and the actors who advocate for and against specific policies. This requires a knowledge of detail, people, and events that is seldom achieved in policy analysis. Using this focused framework in conjunction with the broad policy cycle model we developed our own model, the policy cycle advocacy system, which explains the complexity, variability, and causality in the policymaking process. This book is an attempt to contribute to the theoretical understanding of the policy process. Where does one begin? What does one look for? How should information be categorized?

This book is also the most comprehensive study of ECO to date. Although mandatory ECO programs are not required at this time, the problems targeted by ECO have not been eliminated. We are regularly confronted with information that air quality, traffic congestion, and urban decay are getting worse. In fact, most Americans won't have to read the many studies cited in our book to be convinced; they confront the problem every day during their drive to work.

Since most of us still drive to work alone, a mandatory ECO policy would affect us—we cannot assume the other person will carpool or take the bus to solve our collective problem while we continue to drive alone. Basically, we are running out of options, and therefore we believe we have not seen the last of ECO programs. When we pan across the state and local horizon, our camera finds several recent variations of commute reduction policies, which tells us the problem remains on the government's agenda.

The Policy Cycle Model

At the core of our public policy model, the policy cycle advocacy system, is a conceptual framework that has been used extensively to examine the policy process. The policy cycle model is a useful way to organize our thoughts and helps us understand why policies look the way they do. It envisions the policy process as a cycle that proceeds through a series of stages. Policy analysts have proposed different numbers of stages, ranging anywhere from five to eleven depending on the level of detail desired or the interest of the analyst (O'Connor 1996, Peters 1996, Palumbo 1988, Jones 1977). However, common elements included in all of these manifestations are problem definition, agenda setting, policy formulation, policy implementation, policy evaluation, and policy change. These elements can be viewed as a sequential pattern of activities or functions that can be distinguished analytically, although empirically they may be difficult to isolate (Palumbo 1988). The six stages in the policy cycle are depicted in Figure 1.2.

Figure 1.2   The Policy Cycle

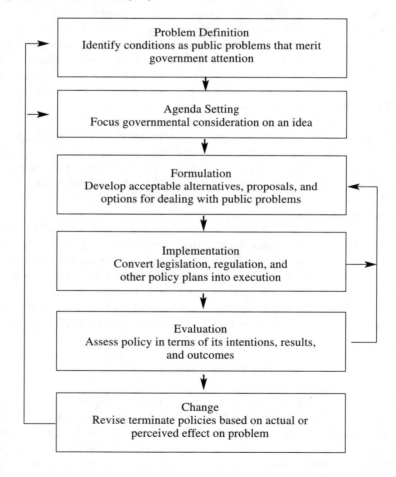

Problem Definition

The function of problem definition is at once to explain, describe, recommend, and above all, persuade. The definition of a problem invariably entails some statement about its origin or cause (Peters 1996). Problem definition focuses on what is identified as a public issue and how different groups within the public think and talk about this issue.

Problem definition has acquired increased importance in the study of policymaking (Rochefort and Cobb 1994). Researchers interested in the appearance of new issues have investigated how the description of a given social problem affects the attention it receives from government. The definition of a problem, with its implicit and explicit causal relationships, often structures the solution. Thus, if air pollution is defined as a problem because there are too many cars on congested streets, then a logical solution to the problem is to reduce the number of cars. However, if air pollution is defined as a problem because of automobile exhaust, then the solution is to reduce tailpipe emissions.

Often problems are not defined initially as having any specific cause. Although a majority may agree that air pollution is a problem, getting a majority to agree on why air pollution is a problem is more difficult. Several groups frequently emerge, each espousing a different definition of the problem. At the nexus of politics and policymaking lies persistent conflict over what causes the problem and, based on the resolution of this conflict, what kinds of solutions should be attempted.

The defining process occurs in a variety of ways, but it always has major import for an issue's political standing and for the design of public solutions. Cultural values, interest group advocacy, scientific information, and professional advice all help shape the definitions of problems. Once crystallized, some definitions will remain long-term fixtures in the policymaking landscape, whereas others may undergo constant revision or be replaced altogether by competing definitions.

Agenda Setting

Before a policy choice can be made by government, a problem must be accepted as a part of the agenda of the policymaking system. In the policy cycle model, the agenda is what the government is doing about problems, or what people think it should be doing. Deciding which problems or issues government should act on is the essence of the agenda-setting process.

Understanding the agenda-setting process is easier if we distinguish between systemic and institutional agendas (Cobb and Elder 1984). The systemic agenda consists of all the issues that some members of the community believe merit public attention and lie within the legitimate jurisdiction of existing governmental authority. "There oughtta be a law!" characterizes the public outcry on these issues. The government has not acted on the problem yet, but some people believe that it should.

The institutional agenda is more specific and concrete than the systemic agenda. It consists of whatever the government is actually doing (or thinking) about an identified problem. Any problem that is explicitly receiving active and serious attention by decisionmakers is on the institutional agenda. Each legislative body in the country has its own institutional agenda; this includes the U.S. Congress (Senate and House of Representatives), the fifty state legislatures, and the numerous city councils and county commissions.

A policy centers not only on facts but also on an interpretation of those facts. The use of symbols is important in moving a problem onto an agenda. Groups try to impose symbols of legitimacy on policies they like by creating catchy phrases like "law and order," "health and safety," and "devolution." These phrases generate emotions that will in turn serve to define the problem and the desired solution in the public's mind. Of course, it helps if there is a triggering event that brings the issue vividly before the public. Triggering events may include natural catastrophes, such as earthquakes, droughts, or record heat, and these also help a problem achieve agenda status.

Problems do not move on and off agendas by themselves. Individuals or groups are involved, and the political conditions must be ripe in order for an issue to get onto the institutional agenda. John Kingdon (1984) calls this a "window of opportunity," which opens for only a brief moment. Similarly, Anthony Downs (1972) suggests that most social problems enjoy a brief "moment in the sun" before public attention turns to newer problems. Although Downs's issue-attention model did prove to be highly predictive of the attention paid to some social problems, issues such as the environment never seem to go off the agenda (Dunlap 1995). Budgetary issues are cyclical by necessity, whereas others, like clean air, are recurrent and suggest a failure of previous policy choices to produce the intended results. Getting new issues on the national agenda or keeping old issues from being taken off is a key activity of policy advocates who coordinate the groups interested in the issue in order to take advantage of any window of opportunity.

Policy Formulation

Policy formulation involves the development of acceptable alternatives or proposed courses of action for solving a public problem (Anderson 1997, 113). This stage of the policy cycle has also been called "policy legitimizing" (Peters 1996). This implies that enacting policy into law gives it a legal and legitimate status. In some ways, this phrase is more accurate than the word *formulation* because elements of a policy are often in operation before they are given formal, legal status. Sometimes policy elements are put into operation at the local level and then adopted and reformulated as national policy. The reverse may also occur when policy elements formulated at a national level must be enacted into law and legitimized at the state and local levels.

Policymakers are often confronted with several competing proposals and may struggle with devising their own alternatives or compromises. Policy formulation does not always culminate in a law, executive order, or administrative rule. Policymakers may decide to take no action on a problem, to leave it alone and let matters work themselves out. Taking no action, often referred to as "nondecisionmaking," is also a formulated policy. In this situation, policymakers consciously decide that they do not want to formally address the problem—at least not at the present time.

Policy Implementation

The most important and critical stage in the policy cycle is the implementation, or execution, of the policy. Without this stage, the policy remains a mere paper plan whose impact on the defined problem will be negligible. Anderson (1997, 214) describes this stage as "what happens after a bill becomes a law." Involving many individuals, organizations, procedures, techniques, and target groups to put the policy into action, implementation also requires organizational capacity to apply the principles of the legislation to specific cases, monitor the performance of the policies, and propose improvements in the formulation and implementation of the policy.

This stage has become even more critical in the 1990s, when more and more state and local governments, the private sector, and individual citizens are required to implement federal policies. ECO policy required four levels of implementation—the federal government to issue guidelines, the state governments to issue regulations, private companies to develop plans, and finally, employees to change their commuting behavior. Each of these four levels was critical to the successful implementation of ECO.

Politicians, bureaucrats, private interest groups, and the general public vie for control of implementation. The intent of a policy may be altered during implementation if the details allow for too many changes and exclusions. For example, although the intent of the ECO program was to reduce the total number of home-to-work vehicle miles traveled during peak commuting periods, the EPA limited the time frame to the morning commute only and focused on vehicle occupancy on arrival at the work site rather than on how many miles were traveled. In their clarification of EPA guidelines, some state regulations permitted companies to exclude from the program disabled individuals or those who drove electric cars. The ECO legislation specifically stated that the program would apply to all companies with more than 100 employees. EPA guidelines, however, interpreted the law to apply to companies with more than 100 employees only if at least thirty-three employees reported to work during a set of hours defined as the morning commute. Thus, a company with more than 100 employees but with less than thirty-three of those employees reporting to work between 6:00 and 10:00 A.M. would not have to implement ECO. The details are indeed very important.

There are five reasons why we consider policymaking to continue during the implementation stage. First, a law cannot always specify the details in advance. In some cases, there are few established techniques for accomplishing policy goals. Laws written to force technology are often amended a few years later after the technology improves.

Second, implementers who do not agree with policy objectives or the methods specified in the laws may resist carrying them out. Resistance may even become sabotage if the new policy upsets established agency routines. We are all familiar with stories of police departments that refuse to enforce nuisance laws—spitting on the sidewalk, loitering, and so on—because they take personnel away from what are defined by the agency as more important duties.

Third, the interorganizational nature of most public policies affects implementation. The need for horizontal coordination becomes a key ingredient of policy success (Peters 1996). Few policies are designed and implemented by a single government organization. Jeffery Pressman and Aaron Wildavsky (1973), who popularized the concept of implementation, describe the problem of implementing policies through a number of organizations as one of multiple clearance points. Each clearance point involves a decisionmaker who must agree with the details before a policy can be translated into action. Difficulties occur in coordinating these decisionmakers at the same organization level. The success of one agency's program may require the cooperation of other organizations. The impact of interorganizational relations is especially evident in social and urban legislation, in which more than one department or agency is involved in putting a single piece of legislation into effect.

Fourth, intergovernmental, or vertical, coordination can be equally complex. Few federal programs are implemented by only one level of government. It is common for the federal government to enact legislation that must then be implemented by states; the states then pass on the implementation to local governments. The discretion and flexibility given to state and local governments will vary depending on the particular problem. Too much discretion can lead to unequal implementation, whereas not enough discretion may result in ineffective implementation. A vertical implementation structure may give rise to inadequate implementation or even no implementation.

Fifth, the public sector's frequent reliance on the private sector to deliver public programs makes policy implementation complex and difficult to coordinate. Activities that once were almost entirely private now have greater public-sector involvement, and vice versa. The push toward privatization of public activities complicates the implementation stage. In implementing public policy, the private sector needs standardized oversight and enforcement by public agencies. If the private sector is to be held accountable for implementation, mechanisms must be established to ensure standardization in compliance. Since federalism makes it difficult to require that all states issue the same regulations, the private sector is affected differently depending on where the company is located.

Policy Evaluation

The fifth stage of the policy cycle involves evaluation of what has occurred. Once they are on the public agenda, public policies are evaluated almost continuously, both formally and informally by citizens, the news media, legislators, academicians, administrative agencies, auditors, and interest groups. Many participants in the evaluation process do not play a formal role, but the persuasive power of these anecdotal evaluations is considerable.

Ideally, a formal evaluation strategy is set up before the policy is implemented. In this way, decisionmakers can specify the expected purpose of the new policy and what changes will indicate success of the new policy. The formal evaluation should be done systematically, using the best research methods available. In reality, evaluations seldom conform to this ideal. As in other stages, policy evaluation has both political and technical components. Anecdotal or informal evaluations based on conversations with a few unhappy participants may be magnified in the media and be the basis for changing or terminating a policy.

Policy evaluation should consider the consequences, both intended and unintended, of the policy. Intended consequences are the goals stated during the formulation of policy, that is, the effects that policymakers explicitly said they wanted from the policy. Policy evaluation asks: How well have these goals been met? But policies also produce unintended consequences. Both unintended consequences and failure to meet explicit goals may force a policy to be redefined (i.e., sent back to the problem definition stage), amended (i.e., returned to the policy formulation stage), or eliminated (i.e., taken off the agenda).

Policy Change or Termination

In an ideal world, termination would be the sixth stage in the policy cycle. Termination means that the problem is solved or significantly ameliorated so that government action is no longer necessary. In the real world, however, problems are not readily solved, and termination is not a typical policy outcome. When government programs are terminated, it is more likely due to political or ideological opposition (informal evaluation) than to the elimination of the problem or the recommendation of a formal evaluation (de Leon 1983).

It is far more typical for a policy to be reformulated, modified, or changed than it is for it to be eliminated (Hogwood and Peters 1985). To begin with, terminating the original investment of resources and commitment is costly. Further, once implemented, programs create stakeholders— people or organizations directly or indirectly involved with or benefiting from the policy who have vested interests in maintaining it. Stakeholders are much more cohesive in fighting against the termination of a program than interest groups or advocacy coalitions are in pressing for its creation. Policy termination has clearly identified losers, whereas in policy formulation, when goals are often stated in general and vague terms, it is not always clear who the winners and losers will be.

Advantages and Disadvantages of the Policy Cycle Model

Conceiving of policymaking as a cycle with successive stages has several advantages. The principal one is that it enables us to envision the entire process from beginning to end. We can therefore conceptualize policy as a whole, not just as a single law or regulation. Laws passed by legislative bodies are only part of policy and, indeed, may be only the smallest part. When a law with vague and ambiguous objectives is passed—which is frequently the case in U.S. politics—then policy really is formed by the administrative agencies who implement the law.

Despite the advantages of this abstract representation of the policy-making process, it has some disadvantages. The model is too neat, too logical, and too sequential (Palumbo 1988). In the real world, the various stages overlap and intermingle, occurring at the same time or out of sequence. It is often difficult to separate problem definition from policy formulation. Policy is always being formed and reformed: it is never a single, clear, and noncontradictory set of objectives. It is made not only by policymakers at the top, but also by the many "street-level bureaucrats who actually deliver services" (Lipsky 1980).

The stages depicted in Figure 1.2 show subloops running from implementation and evaluation back to formulation. These subloops illustrate that the policy cycle is a dynamic and iterative process during which policies are often adjusted based upon knowledge about their actual impact and shortcomings. As programs are applied to different conditions, more is learned about what works: "What the observer sees when he identifies policy at any one point in time is at most a stage or phase in a sequence of events that constitute policy development" (Eulau and Prewitt 1973, 481).

The policy cycle model also lacks a causal mechanism. What or who moves policy from one stage to the next? Why do some policies move fairly quickly through the process, while others linger on the systemic or institutional agenda indefinitely? It seems clear that we need to focus on the individuals and groups who act to identify a problem, consider options, and allocate resources, but nothing in the policy cycle model points us in an explicit direction.

The Policy Cycle Advocacy System

As we examined the ECO provision of the CAAA-90, we became interested not only in the policy itself, and how it changed as it progressed through the policy cycle, but what moved it through the policy cycle. However, the shortcomings of the policy cycle model led us to search for another model that could explain how the policy cycle operates by identifying the causal factors within the cycle. We turned to the advocacy coalition framework, most frequently associated with Paul A. Sabatier (1988) and Sabatier and Hank Jenkins-Smith (1993), which provides clear assumptions about causality and identification of the people and groups involved in policymaking. It includes variables that are not new to the study of policy but combines these variables into a new coherent approach. However, it does not incorporate the stages of the policy cycle model, which have been so useful to others, as well as ourselves, in analyzing public policy.

We concluded that an integration of the policy cycle model and the advocacy coalition framework would be most helpful in explaining policy change over time. We call our integrated model the policy cycle advocacy system. It is depicted in Figure 1.3. We first identify the policy subsystems associated with the policy as well as the stages of the policy cycle. We then analyze the changes within and between each stage by first examining the policy subsystems involved and then by assessing the activities of the advocacy coalitions within each policy subsystem. We view advocacy coalitions as the engines that move policies through the policymaking cycle. Policy can be traced to the values and belief systems of the advocacy coalitions dominant in a particular stage of the policy cycle. Changes in both stable and variable system factors are also important causal factors. It is important to reiterate that the stages of the policymaking cycle overlap. For this reason we have depicted them in Figure 1.3 in a vertical continuum rather than in the traditional boxes used in Figure 1.2. We now turn to an explication of our integrated model.

Policy Subsystems

The ACF is based on the centrality of policy subsystems, "that is, those actors from a variety of public and private organizations who are actively

Figure 1.3   Policy Cycle Advocacy System

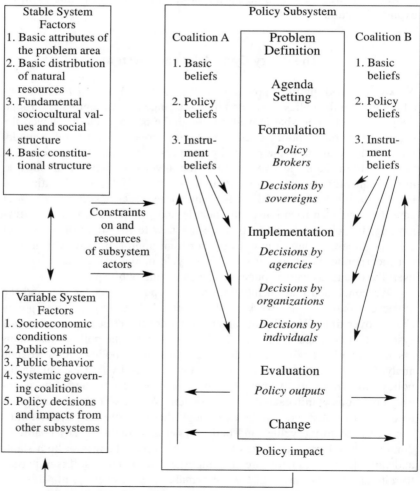

Source: Adapted from Sabatier and Jenkins-Smith (1993).

concerned with a policy problem or issue such as air pollution, health, or surface transportation" (Sabatier 1993, 17). There are multiple policy subsystems within a single political system, and each subsystem is organized according to specific problems or issue areas.

Each subsystem consists of interested actors who seek to influence governmental decisions in a particular area. One of the ACF's four premises is that policy subsystems have an intergovernmental dimension. The actors who compose a policy subsystem come from a wide variety of governmental organizations, including federal agencies, state agencies, Congress, and local governments. Actors within a policy subsystem also include members of corporations and businesses affected by the problem as well as academics, consultants, journalists, and other concerned groups or individuals. Any individual, group, or institution that has an interest in, is affected by, or contributes to a particular problem or issue is viewed as a member of that policy subsystem (Sabatier 1993, 24). According to Sabatier, there are also important latent actors "who would become active if they had the appropriate information" (1993, 24). Activating latent constituencies is a strategy used by coalitions to build support for a new policy or a change in an existing policy. In the case of ECO, companies charged with implementing ECO plans and the employees whose behavior they were supposed to change were latent actors.

Advocacy Coalitions

Since each policy subsystem contains a group of people with diverse perspectives on the problem and solutions, some way is needed to identify fewer and more broadly representative groups. Sabatier calls such groups "advocacy coalitions" (1993, 25). A coalition is made up of individuals "who share a particular belief system, that is, a set of basic values, causal assumptions, and problem perceptions and who show a nontrivial degree of coordinated activity over time" (Sabatier 1993, 25).

Usually only a few (two to four) significant advocacy coalitions exist within a particular policy subsystem.[1] One advocacy coalition tends to dominate policy at a particular point in time, but other coalitions may move into dominance as a result of changes in external constraints and increased ability to attract members and resources.

It is important to emphasize that the term *advocacy coalition* is a theoretical construct rather than any real organization or group. In other words, no one would actually identify himself or herself as a member of an advocacy coalition. Rather, they would more readily identify themselves as members of an organization, institution, corporation, interest group, or target population (those groups or individuals whose behavior has been marked for change). However, this construct is a useful analytical tool for studying public policy because it enables researchers to aggregate individuals and groups with similar policy views into two or three broad categories, which allows the analysis of public policy to focus on important differences.

Public policies and programs are shaped by and reflect belief systems (Sabatier 1993). The individuals, groups, and institutions within an advocacy coalition share a common belief system that they want to see enacted into public policy. Belief systems, or worldviews, include three levels of beliefs:[2]

- Basic beliefs: These are normative views about the nature of the world and humanity's place in that world. They include fundamental values, such as the belief that air pollution is bad, which are very hard to change.
- Policy beliefs: These include problem definitions, causal assumptions, policy positions, and strategies for furthering the basic beliefs. One example is the belief that people driving too much causes air pollution. Policy beliefs are relatively stable but are not as hard to change as basic beliefs.
- Instrument beliefs: These are beliefs about the best methods, or instruments, for achieving the policy beliefs. For example, an instrumental belief that people should be required to drive less is appropriate if the policy belief is that people driving too much causes air pollution. However, that instrument belief is less appropriate if auto exhaust is the problem. Beliefs change more quickly at this level as people learn what will work technologically and what is politically feasible.

Groups often organize around basic beliefs, such as the primacy of human health versus jobs and the value of government versus marketplace solutions. Since these basic beliefs are relatively stable, they often provide the basis for identifying advocacy coalitions. The composition of coalitions is also very stable. Two key actors in the U.S. air pollution subsystem in the 1970s and 1980s, Congressmen Henry Waxman and John Dingell, are still dominant players in the 1990s, as are organizations such as the National Association of Manufacturers, the American Lung Association, Sierra Club, and the EPA.

Policy Brokers

Conflicting beliefs and strategies among various coalitions are normally mediated by a third group of actors, called policy brokers. Their principal concern is to find reasonable compromises that will reduce conflict and move the policy to the next stage. Although the ACF differentiates policy brokers from advocacy coalitions (Sabatier 1993, 18), it also acknowledges: "The distinction between 'advocate' and 'broker'. . . rests on a continuum. Many brokers will have some policy bent, while advocates may show some serious concern with system maintenance" (Sabatier 1993, 27). In order to account for this variation, our model separates brokers from advocacy coalitions. In reality, we believe most brokers are also members of advocacy coalitions or that they are, at least, predisposed to support one advocacy coalition over another. A person or group acts more like a policy broker than an advocate when they show more concern for the stability or maintenance of the political system than the superiority of their belief system.

Decisions by Sovereigns

Decisions by sovereigns occur primarily during the policy formulation and implementation stages. Advocacy coalitions compete in mobilizing support, evidence, and persuasion to influence these decisionmakers. Efforts by competing advocacy coalitions to advance their beliefs can delay decisions for years. Passage of the CAAA-90 took thirteen years and a president, George Bush, committed to getting it through the legislative process. The supremacy of the federal constitution guarantees that laws passed by the federal government will not be overturned by a lower level of government. However, subordinate levels of government have been known to pass laws that inhibit, or even prohibit, the carrying out of a federal sovereign decision. For example, in 1995 the Maryland state legislature passed a budget expressly prohibiting any expenditure of state funds to implement ECO (MDE 1995, 9). The Pennsylvania state legislature engaged in a similar action in 1994.[3]

Decisions by Agencies

According to Sabatier and Jenkins-Smith, "administrative agencies have the most direct impact on a coalition's ability to achieve its policy objectives because agencies are the institutions that actually deliver services or regulate target group behavior" (1993, 227).[4] The resources and general policy orientation of the implementing agencies play an important role in the policy subsystem. "Administrative agencies must not, however, be viewed as monolithic. First, different sections of the same agency may be allied with different coalitions. Second, the official agency position espoused by political appointees may not be followed by subordinates within the agency" (Sabatier and Jenkins-Smith 1993, 214).

Administrative agencies are assigned the task of issuing rules and directives that will fill in the details of policy and make it specific enough to implement. A single government agency is usually assigned this task, but in some cases, the job may be assigned to more than one agency at more than one level of government. In ECO one federal agency was required to issue guidelines instructing state agencies to issue regulations on how to implement the policy. This intergovernmental dimension increases the implementing agency's discretion and latitude.

Decisions by Organizations and Individuals

Organizations and individuals outside government are often neglected by analysts of policy implementation, but their decisions are key to understanding the success of the policy.[5] Most policies identify a target group(s) whose behavior must change if the policy is to have the expected output and impact.[6] Individual "street-level" decisions to comply or not comply with public policy are influenced by the policy itself, advocacy coalitions, and external factors. They are not random.

In some cases, the policy identifies only organizations as its target group. Public and private organizations are required to change their operating procedures (Occupational Safety and Health Act) or purchasing habits (Paperwork Reduction Act). In other cases, the policy identifies individuals as its target group; for example, individual employees are required to don safety gear before operating a machine or are expected to recycle office waste. In still other cases, both are targeted, and the organization is held responsible for changing both its own behaviors as well as the behaviors of its employees. In these cases, organizations are generally permitted to use a variety of tools (incentives and disincentives) to obtain compliance from their employees in the work setting.

In the case of ECO, however, organizations were held responsible for the behavior of their employees outside work but not given any tools to obtain compliance. ECO policy mandated changes in the commuting behavior of individual employees in order to improve the quality of urban air, assuming that employers could change their employees' commuting behavior and paying little attention to whether this was possible or how it could be done. ECO was not fully implemented as planned, but we did learn something about the decisions of organizations to implement policy. Some companies implemented ECO programs, whereas others refused even to discuss ECO. Among companies considering ECO programs, employee surveys provided some insight on the commuting decisions individual employees might make, since employees reported the probability that different types of company ECO programs would affect the way they commuted to work.

Advocacy coalitions can affect policy by influencing the decisions of organizations and individuals through persuasion, education, or direct assistance (Mazmanian and Sabatier 1989). Many times the advocacy coalition supporting a policy loses interest once the policy has been formulated, and at the same time, the interest of an opposing coalition increases and it begins persuading organizations and individuals not to comply. Even when the general public initially supports the policy, public support almost invariably declines over time as new issues capture the public's attention or the costs of the policy become more apparent (Lester and Bowman 1989; Downs 1972).

Policy Outputs

Policy outputs are the things actually done (or not done) by target populations in response to policy. Outputs are usually concrete and can be readily counted, totaled, and statistically analyzed (Anderson 1997, 276). Policy must have outputs, but not all outputs help solve the problem. Sometimes outputs contribute nothing more than "bean counting" (Gormley 1989, 5). Agencies, organizations, and even individuals under pressure to demonstrate results "may focus on outputs in order to generate statistics that create the illusion of progress" (Gormley 1989, 5).

ECO specifically stated that companies with more than 100 employees in the affected nonattainment areas were to increase their average vehicle occupancy by 25 percent. This is a bean counter's dream! The law clearly defined success as an increase in vehicle occupancy, but an increase in vehicle occupancy is not synonymous with a reduction in air pollution. Achieving policy outputs does not always translate into having an impact on the problem.

Policy Impacts

Policy impacts, in contrast to policy outputs, are the consequences for society, intended and unintended, that come from deliberate government policy (Anderson 1997, 276). According to Daniel Mazmanian and Paul Sabatier (1980, 443), a policy must be based on "an adequate causal theory" with a clear linkage between the policy and its objectives. The policy will have no impact on the problem if no causal relationship exists between the policy and the problem. Even when a relationship does exist, support for the policy is diminished when the causal connection is not clearly communicated or is controversial, and consequently the impact of the policy is diminished.

Measuring policy impacts is more difficult than measuring policy output. It is often especially difficult to attribute an improvement in the problem solely to the existence of the policy. For example, in 1997 the skies over Los Angeles were the cleanest since monitoring began four decades ago. Environmentalists argued that years of cutting-edge and costly anti-smog improvements had produced this result. Skeptics contended that the cause was El Niño, a periodic weather phenomenon that helped flush the Los Angeles skies with "nice, fresh, unstable tropical air, which broke up the inversions that normally trap noxious air over the basin" (Booth 1997, A1).

External Constraints

Until this point, our model has been predicated on the assumption that policy is a function internal to the policy subsystem, as advocacy coalitions seek to translate their basic, policy, and instrument beliefs into government programs to solve problems they have defined. But in fact, policy is constrained by the larger social and economic system, and policy change may also result from some "external perturbations, that is, the effects of system-wide events" (Sabatier 1993, 34). These external perturbations affect the constraints and resources of all the actors of a policy subsystem: advocacy coalitions, policy brokers, agencies, organizations, and individuals. As Figure 1.3 indicates, these external perturbations can include "changes in socioeconomic conditions, system-wide governing coalitions, and policy outputs from other subsystems" (Sabatier 1993, 34). They can be grouped into two sets of exogenous (i.e., outside) factors, which we label "stable system factors" and "variable system factors."[7]

Stable system factors are assumed to be relatively constant. These factors set the limits on what governments can and cannot do. These include basic attributes of the problem area, basic distribution of natural resources, basic constitutional structure, and fundamental sociocultural values and social structure. Air pollution policy is strongly affected by the composition of air and the nature of chemicals and other materials released into the air. It is also affected by the geographical contours of the land that define air basins. The political boundaries of cities and states, as well as our country's governing principles, affect what solutions are politically feasible. Beyond this, public policies generate political support when people think the policies will promote their fundamental sociocultural values. They generate opposition when people think the policies conflict with their fundamental values (Sabatier and Mazmanian 1979). However, sociocultural values are seldom explicitly or rationally defined; rather, they are general, amorphous, deep-seated beliefs about such things as freedom, private property, and privacy. Different values held by an individual or group may conflict in specific situations addressed by public policy.

Although sociocultural values have great stability, they can and do change. The U.S. public's view of the environment changed dramatically over 200 years. The dominant sociocultural values in the early days of the republic were essentially antienvironmental. Americans viewed natural resources as unlimited, for people's exclusive use and enjoyment, and needing control through technology. Technology also could fix any ecological problems that might arise. Now Americans largely view the environment as limited, fragile, and in need of protection (Dunlap and Van Liere 1978). These new environmental values conflict with the American value of individual freedom, particularly when freedom is defined by the personal mobility provided by the automobile. ECO policy developed in the midst of these conflicts. Its supporting advocacy coalition thought most Americans, since they embraced environmentalism, would accept commute reduction as a means to reduce air pollution. However, this coalition may have underestimated the value Americans place on individual freedom and mobility.

The second set of external constraints, variable system factors, includes socioeconomic conditions, public opinion, systemic governing coalitions, and policy decisions and impacts from other subsystems. To this list our model adds a fifth factor—public behavior—which we believe is distinct from public opinion and socioeconomic conditions.

Socioeconomic conditions affect how problems are perceived and defined and set the stage for possible solutions. Wealthy countries may have the "luxury" of worrying about environmental pollution, whereas poor countries are too busy meeting the basic survival needs of their citizens. At the same time, wealthier nations or individuals may place greater demands on environmental resources than those with fewer disposable resources. Certainly the increase in the number of cars, which now outnumber drivers (Pisarski 1996), affects people's choices of carpooling or driving alone.

Public opinion, unlike sociocultural values, is what people think about particular issues. It constrains the range of policy strategies available to policymakers and implementers and must be mobilized in support of specific policies. Public opinion is less stable than fundamental sociocultural values and can change quickly, depending on current events, personal circumstances, and even the weather!

Public behaviors are the actions people engage in every day. They can have an impact on public policy at every stage, but especially during policy implementation. What people do in one area of their lives affects what they do in a related area. Whether people currently engage in pro-environmental behavior, such as recycling, may affect whether they will embrace other pro-environmental behaviors, such as taking the bus instead of driving alone.

Systemic governing coalitions affect policy options available to policymakers. The election of a Republican Congress in 1994 had significant effects on ECO. Congressional committees in the House of Representatives, chaired for the first time in forty years by members of a different party, were substantially more sympathetic to the advocacy coalition that opposed ECO.

Policy decisions in one subsystem can place constraints on or promote policy development and implementation in another subsystem. Many times one problem affects multiple subsystems, and public policy can be best understood by considering all the subsystems together. ECO was such a policy. The problem of air pollution produced by cars during the morning rush hours of many urban areas intersected the clean air subsystem, the transportation subsystem, and the urban subsystem. As we tried to understand ECO policy, the intersection of these three policy subsystems became central.

Conclusion

Public policy has an impact on almost every aspect of daily life. Our research initially focused on how public policy affects groups and individuals. However, as the political environment changed, our research changed to focus on the policymaking process itself. We learned that the policy process is complex, and we needed good analytical tools to fully understand it. Current available policy models missed either the big picture or the details and focused either on the structure or the dynamics. We integrated two popular policy models to provide a more complete picture of what happens when governments make policy, to focus on both the structure and the dynamics. We called our model the policy cycle advocacy system.

We then used our integrated model to analyze the policymaking process that produced the ECO policy and the dynamics of the process. ECO received little attention during the formulation of the CAAA-90, but it provided a unique opportunity to study a policy from inception through reformulation. Unlike most policies, which take years or even decades to implement, ECO was put in place in a rather short time. However, the policy quickly changed from mandatory to voluntary. Our attempt to formally evaluate ECO amid all the events surrounding this change provided us with the grist from which we developed an integrated approach to the study of policy analysis.

Notes

1. Sabatier provides an example: "The U.S. air pollution subsystem in the 1970s and 1980s was divided into two rather distinct advocacy coalitions" (1993, 26). One, called the Clean Air Coalition, was dominated by environmental and public health groups and their allies in Congress. The second, called the Economic Feasibility Coalition, was dominated by industry, energy companies, and their allies in Congress. At any particular point in time, each coalition adopts a strategy envisaging one or more institutional innovations that members feel will further their policy objectives.

2. Sabatier (1993, 31) calls this first set of beliefs "deep (normative) core" beliefs. We find it easier to understand this set by referring to it as "basic beliefs." The second set of beliefs is called "near (policy) core," which we relabel "policy beliefs." The third set of beliefs is called "secondary aspects," which we always found confusing. We refer to it as "instrument beliefs" because it clarifies the focus of this level.

3. Act 95 (75 PA S.C. Sec. 4706(I)) of 1994 required that Pennsylvania's governor suspend implementation and enforcement of the employer trip reduction program until March 31, 1995, or until an alternative program with equivalent emission reductions was developed. The legislation also directed that the employer trip reduction program would not be required if the area was reclassified as a "serious" nonattainment area. See, Pennsylvania, Department of Environmental Resources (1995).

4. Sabatier (1993, 18) labels this "Agency Resources and General Policy Orientation" and indicates that the end result of decisions by sovereigns "is one or more governmental programs, which in turn produce policy outputs at the operational level." We have shortened the label in our integrated model (Figure 1.3) to decisions by agencies, which we believe better characterizes the role played by agencies in the implementation of policy decisions.

5. Sabatier (1993, 18) does not explicitly include this level in his ACF. However, we felt that it was important to include this level because many policies "target" nongovernmental organizations in the implementation of government policy. Private companies are often required to implement social policy. We also felt it important to include this level because individuals who are the target of government policy may be directly or indirectly affected by specific advocacy coalitions as well as by external system constraints. The decisions of these organizations and individuals to comply or not comply with public policy will have a direct effect on policy output as well as its impact.

6. In the case of some policies (tax policy), the target group is extremely large and may comprise a majority of the adult population. In the case of other policies (toxic waste disposal), the group is very small and may comprise a half-dozen companies.

7. Sabatier (1993, 18) labels these two sets of factors "relatively stable parameters" and "external system events." We believe our terms are simpler and more descriptive of the components that make up both sets of factors. We have also included an additional factor under the variable system factors, public behavior, which we believe is an important component.

2

Problem Definition:
The Demand for Action

The specific demands resulting from awareness of problems become the "stuff" of government in a democratic society.

—Charles Jones 1977, 26

The Policy Cycle Begins

The policy cycle begins with a problem. In our imperfect world, there are innumerable problems arising from many sources. Some problems are large; some are small. Some affect many people; some affect only a few. Some are endured, but others are intolerable.

Some problems arise from scientific discoveries and inventions. For example, the invention of the internal combustion engine led to a whole new way of life in the United States, called "automobility," giving individuals great personal freedom in their private vehicles. However, this same invention has had many negative impacts on society and the environment, particularly in cities where large concentrations of cars produce dirty air and traffic jams.

Problems also arise from the tension between the desires of individuals and the needs of the community. This tension can be characterized as a conflict over the "commons," natural, social, and economic resources to which everyone has free and equal access. These resources seem unlimited, with everyone able to use as much as they wish. However, no resource is unlimited; and when the resource runs out, everyone loses. The commons isn't really free; it just seems free to the individual.

Garrett Hardin used the pastoral setting in feudal England to illustrate the "tragedy of the commons." Local villagers had free access to open pasture. Each villager could keep as many cattle as he or she wished on the common grassland. As the number of villagers increased, the number of cattle increased, and the carrying capacity of the commons was soon reached and then surpassed. Overgrazing caused the total amount of grass to decline, which meant the amount of grass available for each cow decreased and, in turn, the amount of milk and meat produced by each cow declined. As each animal became less productive, each individual added another cow to the herd in order to make up for the loss. Although the additional cow further reduced the grass available to every other cow, the whole community shared the burden, whereas the individual reaped the advantage. Since the self-interest of all the villagers was to add cattle to their herds, the total grass resources of the commons and the productivity of each cow continued to decline. The only solution was for the villagers to trade their own "short-term self-interest ... [for] the long-term well-being of the community" (Cope 1995, 118). The herder needed to reduce the size of their individual herds so that there would be more grass for all. However, according to Hardin, this "is the tragedy. Each man is locked into a system that compels him to increase his herd without limit—in a world that is limited. Ruin is the destination towards which all men rush, each pursuing his own best interest in a society that believes in the freedom of the commons" (Hardin 1968, 1244).

The commons scenario illustrates the problems that occur when individuals use collectively "owned" natural and man-made resources. Air and roads are such resources. The car provides many benefits to the individual: people are free to live and work wherever they wish; they have the flexibility to come and go as they want; and they can go to their destinations in safety, style, and comfort. However, automobiles use the air, roads, and urban land "commons," generating costs for the community that are not "paid" for by the individual directly.

Traffic congestion and air pollution are two of the costs generated by the use of the car on the modern urban commons. Cars use a lot of land for roads and parking lots, but the average driver views his or her use of this land to be almost free. Yet someone must pay for the cost of the land, whether directly (monetary) or indirectly (opportunity costs). Moreover, since the use of roads is viewed as free, drivers utilize the roads more than if they had to pay for the use directly. Further, the operation of cars produces waste products that are released into the air. As more people use more cars to travel more places every year, congestion and pollution get worse, and the land and air commons becomes crowded and overburdened. Eventually, someone defines the overburdened commons as a problem for the government to solve.

Problem Perception

Problem perception occurs when someone becomes aware of a situation or an event that he or she considers undesirable and in need of change. If no one thinks that a situation or event is a problem, then no solution is needed. The policymaking cycle begins when someone "notices" a problem exists.

How and when an event is perceived as a problem influences how the problem is later defined and, ultimately, acted upon by the government. For example, volcanoes, storms, meteorites, decomposition, and other natural events release particulates and chemicals into the air and produce unpleasant, and sometimes even deadly, air pollutants. Few view these as problems for governments to solve. Sometimes a situation worsens, and more people perceive the worsening situation as a public problem. For example, when the population was small and there was lots of open space, air pollution wasn't perceived to be a problem because the pollution was dispersed over a wide area and did not affect very many people. As the population grew, the sources of pollution increased, and the area for dispersal decreased, more people were adversely affected and perceived the pollution as a public problem.

Perceptions also change over time. During the nineteenth and early twentieth centuries, smoke from the factories of the Industrial Revolution was viewed as a symbol of economic growth and prosperity. However, by the 1950s and 1960s, this perception had changed. Air pollution from factories was perceived as bad for people's health, and the smoke became a symbol of dirty, unhealthy cities.

Likewise, perceptions of car use have changed. People like the convenience and status of private cars, but one of the consequences of driving cars is exhaust from the internal combustion process. When people first began driving cars, automobile exhaust wasn't perceived as much of a problem because the number of cars were few, and the resultant air pollution was readily dissipated in the air commons. However, as more people drove more, there was more automobile exhaust, and the commons became effectively smaller. Automobile exhaust became both more noticeable and more objectionable to more people.

Problem Definition

When people first perceive a problem they often just put up with it, but then someone decides that the situation is unacceptable. "There ought to be a law!" is the outcry at the beginning of the problem definition stage. Before this outcry can be translated into public policy, however, the nature of the problem must be defined and a theory formed about the causes of the problem. Causality specifies who or what is to blame for the problem and, therefore, suggests how to solve it.

Problem definition is a critical stage of the policy cycle because it lays the foundation for all the later stages. An incorrect problem definition can keep it off the governmental agenda or limit the effectiveness of a solution.

The definition of the problem may be incorrect because of gaps in knowledge, multiple causes that confuse the issue, or strong belief systems that limit policy options. Sometimes people generally agree that an issue is a problem but do not know its cause. The details of the problem become more specific as people debate the issue and eventually arrive at an acceptable definition with its assumed causation.

Advocacy Coalitions

Different people experiencing the same situation can arrive at very different definitions of a problem and its causes. People perceive and define problems differently based upon their beliefs, which often lead to conflict over appropriate solutions for dealing with the problems. Once people perceive and define a problem, they try to bring their definition of it to the attention of policymakers, who tend to be more attentive to groups than to individuals. The bigger, wealthier, more organized, and more vocal the group, the more the policymakers pay attention (Schlozman and Tierney, 1986). Interest groups are formal organizations of people who share similar concerns about various issues, define public problems in similar ways, and promote similar policy goals (Truman 1951). Unlike political parties, however, they do not run their own slate of candidates for public office. Interest groups are found at all levels and branches of government. And although not every interest group is active in the policymaking process, all groups are potentially active (latent) if the government tries to pass a law that would change or affect the groups' activities (Lowi and Ginsberg 1996). Interest groups concerned about a particular issue area interact with each other and with other political actors and institutions in what our model defines as a policy subsystem.

Interest groups, individuals, and governmental institutions that share a particular belief system are aggregated in our model into advocacy coalitions. A common belief system influences the coalition's perception of the problems, their causes, and appropriate solutions. Each coalition seeks "to manipulate institutional rules and actors in order to achieve [its own basic and] policy goals" (Sabatier 1993, 36). Each wants its definition of the problem, its view of causality, and its solution adopted by policymakers. Advocacy coalitions influence which perceptions and definitions are heard and gain popularity in the policy cycle and which solutions are proposed and, ultimately, accepted.

The difficulty for policymakers is how to balance the interests of multiple advocacy coalitions. The policymakers are often themselves members of advocacy coalitions and have their own view of the problem. They may think that a problem has a different level of seriousness and, therefore, define a problem differently from others. Since representation is the linkage between the problem and action, it is critical to consider which perceptions and definitions of the problem are represented among policymakers (Jones 1977, 30).

Policy Subsystems

The problem that commute reduction policy addresses overlaps three policy subsystems—clean air, transportation, and urban. Each policy subsystem has its own issues, political actors, advocacy coalitions, and public policies and its own perceptions and definitions of the problem. However, political actors and advocacy coalitions may be active in more than one policy subsystem, and a given problem may affect more than one subsystem. Commute reduction is a problem that is not wholly part of just one subsystem; rather, it lies at the "intersection" among the three. (See Figure 2.1.) The similarities and differences among the subsystems affected how commute reduction was defined as a strategy for air pollution reduction.

None of the three policy subsystems is primarily concerned with both air pollution *and* driving behavior. Each is more interested in other issues. The clean air subsystem deals most directly with the issues of air pollution and air quality. Pollution from both mobile and stationary sources, airborne

Figure 2.1   ECO and Policy Subsystems

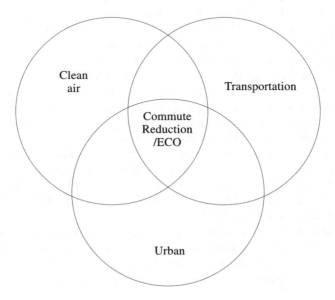

toxins, and ozone depletion are just a few of the issues in the clean air subsystem. The transportation subsystem is concerned with the efficient movement of goods and people around the country. Some issues within this subsystem are interstate traffic (land, water, and air), construction and maintenance of roads and highways, and public transportation systems. The urban subsystem is concerned with the economic development of urban areas, their quality of life, and their use of land. Problems and issues within this subsystem include housing, land use, poverty, environment, public health, and clean air. We turn now to a detailed examination of each of these three policy subsystems and the influence each has had on perceptions of automobile use and the definition of such use as a public problem.

Clean Air Subsystem

Clean air issues cover a wide range of problems—ozone levels, airborne toxins, particulate matter, nitrogen oxide, sulfur oxide, and acid rain. Air pollution as a serious urban problem can be traced back to the Industrial Revolution.

Historical Overview

The Industrial Revolution changed the Western world from a rural, agricultural society into an urban, industrial one. Harnessing energy to power machinery was the dominant feature of the Industrial Revolution, which began in Great Britain in the mid-1700s and spread to other parts of Europe and North America in the early 1800s. Machines were first used on farms. New agricultural inventions such as the reaper allowed one person to do the work of several. Mechanization and innovations in agriculture pushed people from the farms to the cities, where the techniques of the Industrial Revolution were also applied to manufacturing. Machines and assembly lines enabled each worker to produce more goods faster than by hand, which, in turn, enabled businesses to sell more goods at lower prices.

As more sectors of the economy became mechanized, more factories sprang up in the cities. New jobs became available for the farmworkers displaced by agricultural machines. The result was a concentration of people in urban areas. In 1890, 65 percent of the U.S. population lived in rural areas. This declined to 49 percent in 1920 and 20 percent in 1990 (U.S. Department of Commerce 1995).

Industrialization provided inexpensive goods and many jobs, but it also generated unintended and undesirable consequences. A major environmental consequence of the Industrial Revolution was air pollution from factories that vented their waste products into the air. Machines were first powered by steam produced by burning coal. Burning coal produces small particulate matter and smoke, which can hang over a city for days and damages and irritates eyes, throats, and lungs. Burning coal also produces sulfur and nitrogen oxides, which are transformed in the atmosphere into sulfuric and nitric acids and fall from the sky as acid rain. Acid rain is corrosive enough to have very negative effects on buildings, the environment, and human health.

The air pollution produced by the factories and power plants was annoying, but it was not an issue that public policymakers needed to solve. The wealthy, who could afford it, simply moved away from the dirty cities. The poor, especially the factory workers, stayed in the city. By the mid–twentieth century, as more sectors of the economy industrialized and more people "plugged in" to electricity, the number of sources of air pollution grew. As the problem worsened, many people began to perceive this air pollution as a matter for public concern and government solutions.

Homes, factories, and power plants are good candidates for clean-up, primarily because they are stationary. However, in the late twentieth century, a mobile source of air pollution emerged, the automobile. The Industrial Revolution made the automobile affordable. Once a person owned a car, individual mobility increased dramatically, providing a personal solution to the problem of urban air pollution. People could simply move away from the city, live in the cleaner countryside, and drive into the "dirty" city to factory jobs. However, as more and more people used the car to escape the city, and thus drove more miles to work, the air pollution from this mobile source increased.

Air pollution "occurs when gases and particles are combined or altered in such a way that they degrade the air and form substances that are harmful to humans, animals, and other living things" (Bryner 1993, 41). However, many of the compounds that are called air pollutants occur naturally in the earth's atmosphere. They become air pollutants only when someone perceives that they are in the wrong place at the wrong time. For example, ozone in the upper atmosphere is essential for protecting life on earth, but too much ozone in the troposphere is a threat to life.[1] The classification of a chemical compound as an air pollutant is dependent upon another factor as well—time for dispersal: "As long as a chemical is transported away or degraded rapidly relative to its rate of production, there is no pollution problem" (Bryner 1993, 42).

The most common man-made air pollutants are carbon monoxide, ozone, nitrogen oxides, sulfur oxides, and particulate matter. Carbon monoxide is formed by the incomplete combustion of fossil fuels and comes primarily from mobile sources, such as cars and trucks. Ozone is produced in the lower atmosphere when sunlight triggers chemical reactions between volatile organic compounds, nitrogen oxides, and naturally occurring gases. Volatile organic compounds come from mobile sources as well as organic solvent evaporation, which comes from stationary sources during operations in such places as dry cleaners, print shops, and paint shops. Nitrogen oxides are produced by both mobile and stationary sources. Sulfur oxides are generated mainly by stationary sources such as coal-fired electric utilities. Particulate matter consists of tiny fragments of liquid or solid matter floating in the air and is produced by both stationary and mobile sources, such as industrial processes; power plants; cars and trucks; and dust from roads, construction activities, and farming (Bryner 1993).

After the enactment of strict federal, state, and local clean air laws focusing on technological improvements in the automobile, air pollution from mobile sources has shown a steady decline since 1970 (see Table 2.1). The most dramatic decline was in lead emissions. Volatile organic compounds, the primary contributor to ground-level ozone, more than doubled between 1940 and 1970 but declined by nearly 50 percent between 1970 and 1990. However, emissions of other mobile source air pollutants did not decline after 1970. In fact, many Americans (74.4 million in 1990) still live in areas that do not meet the National Ambient Air Quality Standards set in the Clean Air Act Amendments of 1970. The vast majority of these people live in urban areas with high levels of ozone, carbon monoxide, and particulate matter.

Table 2.1   Emissions of Major Pollutants by Highway Vehicles, 1970–1993 (in million short tons)

Pollutant	1970	1980	1990	1993
Volatile organic compounds	12,219	10,990	6,854	6,094
Nitrogen oxide	7,427	8,705	7,488	7,439
Carbon monoxide	79,258	87,991	62,858	59,989
Lead	171,961	62,189	1,690	1,383
Particulate matter PM-10	0.271	0.265	0.239	0.197
Sulfur dioxide	0.279	0.458	0.480	0.438

Source: Adapted from Dorgan 1995, 19–28.

Environmental Opinions

Sociocultural values usually remain fairly stable over time, but Americans' view of the environment has changed dramatically. In the early days of the republic, the dominant sociocultural values were essentially antienvironmental and protechnology. These values had changed by the late 1960s, greatly assisted by Rachel Carson's book *Silent Spring* (1962). Riley Dunlap and Kenneth Van Liere (1978, 1984) found that Americans' fundamental belief system had moved away from an antienvironment social paradigm toward a new environmental paradigm (NEP) that views the biosphere as limited, fragile, and in need of protection. Later research found that the NEP had two (Noe and Snow 1990) and sometimes three dimensions (Albrecht et al. 1982; Geller and Lasley 1985). Our survey of 5,000 employees among fifteen companies found two dimensions. (See Study Participation: Employees in the Appendix for details.) We called the first dimension "Environmental Sensitivity," and it consists of the first ten items in Figure 2.2.[2] The second dimension, called "Man over Nature," comprises the remaining four statements. (See Analytic Measures: Employees in the Appendix for details on the scales.)

Figure 2.2   Environmental Attitudes

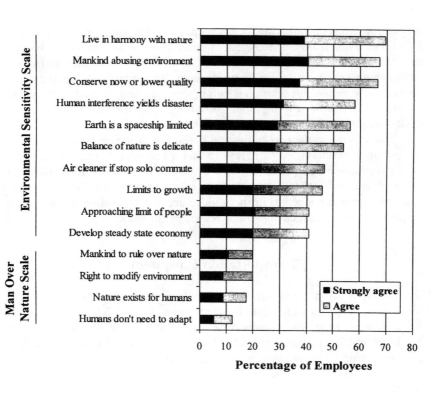

Most employees express attitudes indicating that they are sensitive to the natural environment. Two-thirds agreed or strongly agreed that "humans must live in harmony with nature in order to survive," that "mankind is severely abusing the environment," and "unless we act now to conserve resources, we will soon be faced with a lower quality of life."

Past research found that environmental opinions are largely independent of socioeconomic characteristics (Samdahl and Robertson 1989). Our study confirmed this, although age and family income did have some impact on environmental sensitivity. Employees' sensitivity to the environment increased with their age but decreased with higher family income. With regard to the second dimension, men were more likely than women to believe humans can and should control the environment. (See Table A.3 for the supporting regressions.)

Problem Perception

Regarding urban air pollution, one perception is that growing use of the car will offset gains from technological improvements in automobile emissions and thus behavioral lifestyle changes must occur before air pollution declines any further. Public opinion generally supports this perception. Increasingly, urban air pollution is being perceived and defined as the result of too many people driving alone too often, instead of the sole result of badly designed and poorly operating vehicles.

This change in the perception and definition of the continuing national air pollution problem led Congress to add a behavioral component, commute reduction, to federal clean air policy. In the Clean Air Act Amendments of 1990 (CAAA-90), Congress decreed that a change in driving behavior was needed to further reduce urban air pollution and that consequently people must change the way they get to work.

Transportation Subsystem

The transportation subsystem is primarily concerned with the efficient movement of goods and people by land, water, and air. The transportation subsystem includes the construction and maintenance of transportation facilities, traffic management, and public transportation. Most of the policy activity within this subsystem pertains to the automobile.

The Car

Just as technology spurred the Industrial Revolution from the early nineteenth century, it spurred another "revolution" in the late nineteenth and twentieth centuries. The invention of the internal combustion engine in 1860 made the modern automobile possible. Before the Industrial Revolution, people were very limited in the places they could go. Trains extended travel, although people were restricted by the times and routes of the railroads. The self-propelled personal road vehicle released individuals from the limits of distance, time, and destination (Wright 1992). The first vehicles (1770), powered by steam, were hard to start and dangerous to operate. The internal combustion engine provided a safe and efficient power source for a personal vehicle and was soon followed by other inventions: gasoline engine (1885), electric starter (1912), balloon tires (1922), automatic transmission (1939), air conditioning (1939), and power steering (1950).

The discovery of rich oil fields in eastern Texas in 1901 increased the supply and lowered the price of gasoline. The introduction of assembly line mass production increased the supply and lowered the price of cars, making them affordable for the average middle-class family of the early 1900s. The automobile played its first important role in U.S. society in World War I. Cars were used to transport troops and supplies to the front lines in Europe in record time. Tanks, which were then simply large armored cars, were first used by the Allies in 1916. By the end of the war, the car was a proven, important, and familiar member of modern society. The Great Depression constrained the growth in the number of automobiles, but the economic recovery following World War II spurred the proliferation of cars. By mid-century, the car was firmly established as the preferred way to travel in the United States.

The Roads

However, good roads were needed before automobiles could dominate U.S. transportation. Most roads at the turn of the century were nothing more than wide dirt paths. After a rain, they became virtually impassable for automobiles. "Get a horse!" was a common derisive cry heard by motorists stranded in muddy roads. Without good usable roads connecting the cities and villages in the United States, the automobile would never have replaced the horse and buggy as the most reliable and preferred means of transportation.

During the 1920s, a nationwide campaign was launched to build a network of roads and highways across the country. This effort, interrupted by the Great Depression and World War II, was championed by President Dwight D. Eisenhower for the expressed purpose of defense mobility. The federal government gave billions of dollars to the states to be used exclusively for roads between cities and states, often funding 80 to 90 percent of construction costs. The result is the modern U.S. interstate highway system.[3] In 1921, the United States had only 387,000 miles of surfaced, or paved, roads. By 1940, there were 1,340,000 miles of surfaced roads. Today, there are just under 4 million miles of surfaced roads in the United States, including interstate highways and state, county, and local roads (U.S. Department of Commerce 1998, 627).

Automobility

With cheap gasoline, affordable cars, and good roads, people were ready to drive into the modern auto age. The automobile has produced "a unique way of life based upon personal mobility" (Coughlin 1994, 139). This way of life, often called "automobility," places a social premium on "the ability to go from point A to point B, door to door, at any time, in the comfort, security, and convenience of one's own car" (Coughlin 1994, 139). Individual freedom, one of the most highly cherished U.S. values, is reflected in this automobility. The modern United States is quite simply the most "mobile" country in history.

Automobility changed where Americans lived, how they lived, and where they worked. Before the car, people either lived and worked on the farm or lived and worked in the city. After the car, people no longer had to live near their places of employment. They could now live and work wherever they wished, even if home and work were miles apart. The car is also a place for many activities other than mere driving. With cellular phones and portable fax machines, people can work in the car. They can shop and do their banking in the car. They can listen to instructional tapes and attend drive-in churches. Of course, the car has been used as a pleasure vehicle from the very start. The car has truly become an alternate living and working place and, increasingly, a solo place.

The number of licensed drivers increased by 50 percent between 1970 and 1990, double the growth of the population as a whole. The increase in the number of registered motor vehicles (78 percent) was even greater than the number of licensed drivers, and the number of vehicle miles traveled almost doubled. Figure 2.3 clearly shows that more people are driving more cars (usually alone) and driving those cars more frequently and farther. In contrast to this dramatic increase in demand for roads is a negligible increase in the supply of streets and roads and an actual 2 percent decrease in expenditures for roads during this twenty-year period.

Problem Perception

The transportation picture in the contemporary United States is "more Americans . . . going [alone] to more places more often . . . using their automobiles" (Coughlin 1994, 140). Those in the transportation subsystem

Figure 2.3   Change in Automobility Indicators

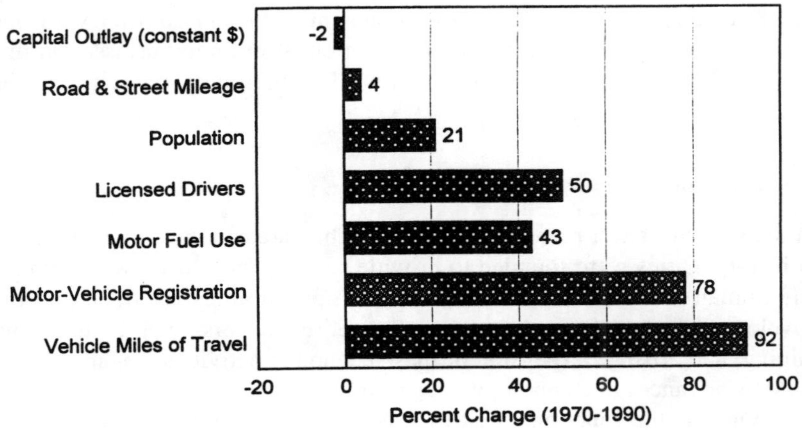

Bar chart showing Percent Change (1970–1990):
- Capital Outlay (constant $): -2
- Road & Street Mileage: 4
- Population: 21
- Licensed Drivers: 50
- Motor Fuel Use: 43
- Motor-Vehicle Registration: 78
- Vehicle Miles of Travel: 92

Source: Federal Highway Administration (1992, 6).

must decide whether this picture is problematic. Some answer: "Yes, it's a problem, but it's not so bad. We can adjust to it and live with it. The supply of roads and technology have not kept up with our needs, but we will just build more and better roads and cars." For others, Americans' love affair with the car is not just a problem; it is a *tragedy* of the urban commons. These people perceive that automobile demand has exceeded the supply of roads in most metropolitan areas.

Until recently, the definition of the problem centered on the technological side of the issues. The solutions proposed have been technological—constructing more and better roads, developing better traffic control measures, or building bigger parking facilities. However, these solutions have not solved the problem. As people drive their cars more and farther, new roads and parking lots quickly become as congested as the old ones. As a result, a new definition is emerging that says that there are too many cars on the road and the solution to the problem requires behavioral changes, getting people to drive less and reducing solo driving.

Urban Subsystem

The urban subsystem's primary concern is the welfare of city residents. Its main policy goal is to establish an economic engine that provides the jobs, necessities, and amenities that contribute to a good quality of life. Two important issues in this policy subsystem are the "economic health" and the "environmental health" of the city. Economic health governs the jobs and money available to city residents. Environmental health affects the quality of life of city residents. These two issues are often in conflict with one another in urban areas. Although air pollution is an important issue in this policy subsystem, it must compete with all of the other issues facing the city.

Urbanization

In the United States, more people live in urban areas now than ever before in history. Cities were founded to provide security, both from wild animals and human enemies. Cities also provided opportunities for trade and the development of crafts and other specialized endeavors, such as tailoring, mining, and artistry. Expansion of these economic activities was made possible by advances in technology in agriculture.

Whereas the function of cities was shaped by economic and security concerns, the shape of cities has been largely defined by transportation issues. Business and industry need to be at the center of transportation routes. People without private means of transportation must live within walking distance of their jobs or public transit. In early industrial cities, the working poor generally lived in the center of the city, the "dirty" areas. The wealthier merchants and factory owners lived on the outskirts, the "clean" areas, using horse-drawn carriages to get back and forth to the central city. This reversed the housing pattern of preindustrial cities, in which the rich lived inside the city, where they were safest from enemies, and the poor lived on the riskier outskirts.

Suburbanization

In contrast with preindustrial cities, the modern industrial city is very decentralized, as people live farther and farther away from each other and their work places. Public transportation gave urbanites a chance "to settle... beyond the walls of the citadel," to live in the country (Bollens and Schmandt 1975, 12). It transported city dwellers from inner-city factories to outer-city homes.

The car was just one factor in the suburbanization of the city after World War II. A second factor, loans offered by the Federal Housing Administration (FHA) and Veterans Administration (VA), enabled more low- and middle-income people to buy their own homes. Third, postwar economic prosperity meant that people were much better off financially than they had been before the war and were able to buy automobiles and homes. The fourth factor was the expanding construction industry, which boomed after the war. Developers bought cheap land on the outskirts of cities and built thousands of suburban "tract" houses. The automobile made it possible for the newly prosperous city workers to live in their inexpensive suburban homes, purchased with government loans, and drive to their city jobs on the new highways. Further, these amenities were available to a larger proportion of society.

Retail activities quickly followed the migration of homes from the central city to the suburbs. Suburban malls replaced the central business district as families' primary shopping locations, and jobs soon migrated to the suburbs. Factories and businesses followed their workers out of the city and benefited from the untapped labor pool of highly educated housewives willing to work for low wages (Garreau 1991). This suburban female labor force needed their own cars to get to their jobs. Today people can live, shop, and work in the suburbs without ever going into the central city.

Automobiles have truly changed the face of the United States. Nearly half of the urban space of the average U.S. city is devoted to the automobile: to roads, parking lots, repair garages, body shops, auto stores, car washes, and the other myriad elements of the automobile infrastructure. In Los Angeles, two-thirds of the land is devoted to the car (Renner 1989, 39). The car even has its own room in the modern U.S. home! According to Charles Jones, not only our space but our time is preoccupied with the car: "We build [and repair] roads, repair the automobiles, take care of the residue, feed the creatures, treat the polluted air, control the traffic, bury the dead, pave the central business districts for parking, spread [millions of tons of] salt on highways in the winter, etc., etc." (Jones 1977, 21–22). Quite simply, cars dominate American life.

Problem Perception

Urbanization and suburbanization occurred as people sought a better quality of life. The accompanying fundamental changes in residence and employment patterns are now at the heart of many of the problems threatening that very quality of life in the urban United States. Both the concentration of economic and population growth in urban areas and the decentralization of this growth into extensive suburbs have resulted from and now require the automobile. The growth of automobile use is overwhelming the ability of both the public sector and the natural ecosystem to respond.

Some individuals perceive the problem as a technological one, for example, inefficient and insufficient roads, and advocate the development of new roads and new technology. Others see a behavioral problem and define it as too much driving. Their solution is that only a reduction in driving, especially solo driving, can alleviate air pollution and traffic congestion.

Complicating this situation is the fact that most modern urban areas extend beyond the political boundaries of "cities." The U.S. Bureau of the Census used "metropolitan area" in 1910 to describe the functional urban area surrounding a central city or cities. The Census Bureau revised the term in 1950 to "standard metropolitan statistical area" (SMSA) (Stark 1992, 569–570). The lack of a single governmental entity for the entire urban area makes it very difficult for policymakers to deal with urban problems, such as congestion and air pollution, caused by the increased use of the car in a sprawling suburban landscape.

ECO: The Intersection of Policy Subsystems

In all three policy subsystems the automobile is perceived as part of an ongoing problem, whether air pollution, traffic congestion, or suburban sprawl. Although technological solutions have been tried in the past, the definition of the problem has changed to include a behavioral component.

The employee commute option (ECO) program intersected the clean air, transportation, and urban subsystems by linking air pollution to the use of the automobile in the city. Initially, only political actors in the clean air subsystem defined solo automobile use as a problem. Political actors from the other policy subsystems got involved in the commute reduction debate only when ECO linked air pollution to traffic congestion and urban sprawl.

The Commuting Problem

The roads and public transportation systems of early twentieth-century U.S. cities were designed to facilitate travel to and from the central city. However, the flow of commuters from suburb to suburb is now twice that from suburbs to city. In 1980, 12.7 million workers nationwide commuted from the suburbs to the central city, whereas 25.3 million workers commuted within and between suburbs. At the same time, 20.9 million workers commuted within the central city, and 4.2 million reversed the commute, going from the central city to the suburbs (Pickrell 1985; Cervero 1989; and Jensen 1993).

In all these work trips, the solo commuter in his or her own private vehicle accounted for 86 percent of work travel in 1980, up from 70 percent in 1960. Only 10 percent of U.S. workers drove with someone else, and 13 percent took public transit or taxis. The remainder walked, bicycled, or worked at home (Pisarski 1987). The vast majority of U.S. households had two or more vehicles in 1980, and the more workers in the household, the more vehicles the household had.

Driving alone was so important to employees in our study that 30 percent said they would drive alone even if they were charged $10 per day for parking. An additional 15 percent said they would switch to carpools only if daily parking charges rose to $6–10 per day. When directly asked their opinions, two-fifths of the employees said carpooling was unpleasant and that it was not something they talked much about. Further, most employees said it was unlikely that they would carpool in the future. Employees' environmental attitudes, especially environmental sensitivity, were the best predictors of employees' attitudes toward carpooling, better than all of the socioeconomic characteristics of employees combined. The more sensitive employees were toward the environment, the more favorably they viewed carpooling. (See the Table A.3 for the supporting regression.)

Seven of ten employees in our study drove alone to work every day during the week. (See Figure 2.4.) An additional one out of thirteen drove alone sometime during the week. One employee in six carpooled regularly, most frequently sharing a ride with one other person. Only one employee in sixteen walked, bicycled, or used public transportation every day. (See Analytic Measures: Employees in the Appendix for a description of the average daily car contribution and Table A.3 for the supporting regression.) The number of cars per household was the key predictor of how an employee commuted to work. The greater the availability of a car, the more likely the employee was to drive alone to work.

Solo commuting takes less time and provides the luxuries of privacy and convenience. Solo commuters travel an average of 10 miles in nineteen minutes, whereas ride sharers travel an average of 14 miles in twenty-five minutes. Public transportation commuters (bus, streetcar, subway) travel

Figure 2.4   Employee Commute Mode

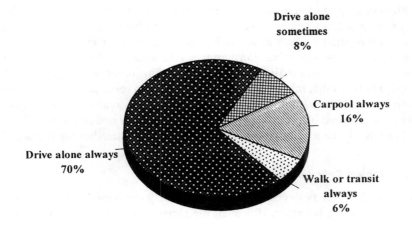

Drive alone
sometimes
8%

Carpool always
16%

Drive alone always
70%

Walk or transit
always
6%

the same distance as solo commuters (10 miles), but it takes them an average of 42 minutes (Pisarski 1987). Solo commuting turns the car into a "time machine" (Garreau 1991). It involves fewer frustrations with timetables and schedules, less perceived danger than public transportation and walking, and a greater sense of being in control.

Americans greatly value the freedom and privacy that comes with driving alone and are willing to accept the costs that come with it. The car is viewed as private space, like a home, and the time spent in it is jealously guarded. Solo commuting is also encouraged by the abundance of free parking in suburban activity centers, the absence of high occupancy vehicle highway lanes, and the high costs of providing convenient, low-cost public transportation in low-density suburbs.

Advocacy Coalitions

One definition of the commuting problem includes a causal connection between increased automobile use and decreased quality of life. According to this definition, many urban problems, not just air pollution but traffic congestion and urban sprawl as well, result from the overuse of the car. As more people live in urban areas, the quality of life decreases, more air pollution is produced, congestion gets worse, the urban area continues to get bigger and bigger, and more and more land is paved to accommodate the automobile. The increasing concentration of jobs, people, and cars in metropolitan areas is simply overcoming technological solutions to the problems. A more effective solution proposed by those who hold this view is to change driving behavior—people must drive less and drive alone less. Our survey of employees found that half agreed that "the air of our cities would be cleaner if people stopped driving alone to work." (See Figure 2.2.)

Others do not share this definition of the problem in U.S. cities. Although air pollution, road congestion, and urban sprawl are serious problems, their solutions do not require that people give up driving their cars. They define the problems as technological ones that cleaner cars and more and better roads will solve. Furthermore, advocates of this definition are very concerned with maintaining and increasing urban employment and oppose policies that might jeopardize jobs for urban residents. According to their perception, restricting the use of the car in the urban area means giving up a key need of the urban economy—the mobility and flexibility of the urban labor force to get to the work sites where and when they are needed.

ECO Problem Definition

None of the three policy subsystems is primarily concerned with air pollution and driving behavior. However, the clean air subsystem, which deals directly with issues of air pollution, is closely connected to the larger environmental movement. Environmentalists have been able to convert favorable public opinion into public action supportive of the environment. Studies show a linkage between environmental attitudes and environmental behaviors (Newhouse 1990; Sivek and Ford 1990; Arbuthnot 1977; Samuelson and Biek 1991). Although there is little research linking commuting attitudes and behavior with environmental attitudes and behavior, advocates in the clean air subsystem reasonably expected the relationships between attitudes and behaviors shown in Figure 2.5.

Figure 2.5   Hypothesized Causal Relationships

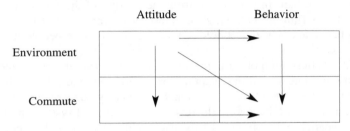

The top horizontal arrow represents the assumption that environmental attitudes affect environmental behavior. Survey data from about 5,000 employees in our study support this assumption with a relatively strong standardized multiple regression coefficient (beta = 0.27). (See Table A.3 for the supporting regression.) The bottom horizontal arrow indicates that commuting attitudes affect commuting behavior. Again, our employee data support this assumption with a relatively strong relationship (beta = 0.23). Defining cars as a key contributor to air pollution implicitly links driving with the environment and assumes that strong environmental attitudes will affect attitudes toward commuting. This assumption is shown by the left-hand downward arrow and was also supported by our employee data, although it was not as strong a relationship as the horizontal ones (beta = 0.19). Likewise, general environmentally responsible behaviors are assumed to affect the environmentally responsible behaviors of carpooling, taking transit, or walking to work, an assumption captured by the right downward arrow. This assumption was not supported by our employee data, however. Those engaging in environmentally responsible behavior in other areas were less likely, not more likely, to use carpools, to walk, or to take public transit as alternatives to driving alone (beta = -0.08). Finally, if driving alone harms the environment, commuting behavior should be directly affected by environmental attitudes, shown by the diagonal arrow in the figure. Again, our data did not support this assumption. Employee environmental attitudes had no relationship to their commuting behavior. (See Table A.3 for the supporting regression.)

Conclusion

ECO illustrates that how a problem is defined structures the solutions considered by policymakers. The old definition of air pollution as a technological problem has been replaced by a behavioral one. According to this new definition, there is a causal connection between increased automobile use and decreased quality of life and that many problems—air pollution, traffic congestion, urban sprawl—result from the overuse of the car. Others do not share this definition, believing that these problems can still be solved by using better and newer technology that does not require people to give up driving cars.

ECO's definition of air pollution linked dirty air to the use of the car in the city and intersected the clean air, transportation, and urban policy subsystems. None of the three subsystems is primarily concerned with air pollution *and* driving behavior. However, each of the subsystems brought its own concerns, political actors, advocacy coalitions, and public policies to the commute reduction debate. Each has different assumptions about the causal connections between cars and the problems in their area of concern, but each defines the problems as worst during rush hour when many people drive alone.

The intersection of subsystems complicates the problem definition stage by bringing together multiple perceptions, definitions, and causal assumptions about the various problems. Many times, "fights" in the problem definition stage occur over these differences, which are based upon the belief systems of the active advocacy coalitions. Which advocacy coalition "wins" the fight in the problem definition stage and gets its perception and definition of the problem accepted by policymakers decides which issues make it to the agenda setting stage of the policy cycle where various solutions to public problems are debated and discussed.

Notes

1. Scientists divide the earth's atmosphere into four spheres: troposphere, stratosphere, mesosphere, and thermosphere. The troposphere is the only layer that can support life and extends 6 to 10 miles above the ground. The air of the troposphere is a mixture of nitrogen (78 percent), oxygen (21 percent), argon (0.9 percent), carbon dioxide (0.03 percent), and trace amounts of neon, helium, krypton, hydrogen, xenon, methane, and nitrous oxide. Some of these gases combine to form

other compounds, such as water vapor, sulfur dioxide, and nitrogen dioxide. Biological and geological processes (for example, decomposition and volcanic eruptions) produce trace amounts of other gases, such as ammonia, methane, hydrogen sulfide, carbon monoxide, and sulfur dioxide.

2. This scale combines Albrecht et al.'s (1982) "Balance of Nature" and "Limits to Growth" scales.

3. President Dwight D. Eisenhower was a key figure in the campaign for an interstate highway system. After World War II, he commissioned a cross-country trip with military vehicles to investigate the adequacy of the nation's roads for military use. He wanted to know if, in a military emergency, troops could be easily transported around the country via automobile. He found the roads to be sadly inadequate. During his administration, he pressed hard for an interstate highway system that would link all the major cities in the United States, making not only military but civilian transportation easier and safer. His vision and efforts were commemorated with the naming of Interstate 70 as the Eisenhower Highway. These interstate highways greatly facilitate travel between cities (Wright 1992).

3

Agenda Setting:
Taking the Demand to Government

The demands that policy-makers choose to or feel compelled to act on at a given time, or at least appear to be acting on, constitute the policy agenda.

—James E. Anderson 1997, 99

The Policy Cycle Advocacy System

The policy cycle deals only with public problems. Some issues are considered private problems and not appropriate for public action. Other issues have long been viewed as public problems and the legitimate responsibility of government. Agreement about which issues are private and which are public changes as circumstances and people change. Once an issue is perceived and defined by some people as a public problem, they have to bring it to the attention of the government. Agenda setting, the second stage of the policy cycle, involves deciding which problems should receive governmental attention.

Our policymaking model distinguishes between two agendas. The systemic agenda includes the issues some of the general public think the government should do something about. The institutional agenda includes those issues that the government actually is considering (Cobb and Elder 1972). The first is the call for action, the discussion agenda, when issues are publicly raised and discussed. The second is the action government takes. This action agenda may set into motion the passage of laws, the promulgation of rules, and the development of programs (Anderson 1997, 99–100).

The Systemic Agenda

Today, it seems that every issue is on the systemic agenda. People talk about many issues and seem to want the government to do something about most of them. In the past, some issues were outside the scope of public discussion and considered private matters for individuals and families. Now it is hard to find exclusively private matters. More and more problems are being publicly discussed and inevitably someone wants the government to get involved. The systemic agenda consists of all the issues that an individual or a group thinks deserve governmental attention (Cobb and Elder 1984). The government may not have acted on the problem yet, but some people believe that it should.

A problem moves onto the systemic agenda after it has been perceived and defined as a problem. Those individuals and groups most interested in the problem begin to discuss the problem as something within the legitimate jurisdiction of government (Cobb and Elder 1984). However, many issues that get on the systemic agenda are discussed and debated, but nothing of real substance is done about them. Avoiding the issue, deciding to do nothing, is a course of action often taken by policymakers. Policy analysts seek to discover what moves issues from a public outcry to official acceptance of the problem, from the systemic to the institutional agenda.

Advocacy coalitions play key roles in moving issues from the systemic to the institutional agenda. They educate the public, media, and policymakers about their perception of the problem and its solution to stir up interest, enthusiasm, and support for the issue. They are successful if they can convince enough people, or at least the "right" people, that the problem merits governmental attention.

The Institutional Agenda

The institutional agenda is whatever the government is actually doing about an identified problem. Governments can do many things to deal with a problem. They can research the matter, analyze plans, submit proposals, debate policies, enact laws, write regulations, authorize spending, enforce rules, and evaluate programs. All these actions constitute the institutional agenda. All are legitimate actions of authoritative policymakers seeking to address public problems (Cobb and Elder 1984).

Many institutional agendas exist in the United States. All the activities previously mentioned occur at every level of government—local, state, and national. A good deal of overlap occurs among these agendas, with some issues appearing simultaneously on several agendas at once. Each level of government has different jurisdictional authority, different internal resources (spending authority, budget, and personnel), different policy concerns, and different political forces at work. The intergovernmental nature of institutional agendas produces a tension in the policymaking process.

Public participation in the affairs of government is a hallmark of a representative democracy. In setting the institutional agenda, policymakers must determine which issues citizens think should receive governmental attention. However, no citizen can be fully aware of all of society's problems, and citizens who are aware may not be able to influence the government directly. Public officials are simply less responsive to the complaints and petitions of a few citizens than they are to those of many citizens, particularly when they organize.

Numbers alone are not enough, however. Interested individuals and groups must persuade at least one public official that enough people, especially those important to the official, identify the issue as a public problem that cannot be ignored. Generally this persuasion comes from advocacy coalitions of diverse individuals, interest groups, private institutions, and public institutions that are concerned about similar problems. Members of these advocacy coalitions have similar belief systems about the world, its problems, and their solutions. They cooperate in significant activities directed toward getting their problems, definitions, and solutions on the institutional agenda so that their beliefs are enshrined in public policy (Sabatier 1993, 26).

Advocacy Coalitions

The belief systems of advocacy coalitions consist of basic beliefs about fundamental values, policy beliefs about the nature of these problems and the best strategies for dealing with them, and instrument beliefs about the specific methods governments should use to resolve these problems.

An advocacy coalition's belief system determines the direction it wishes public policy to go, but its resources determine its ability to influence policy. These resources include money, expertise, the number of members, legal authority, media attention, and grassroots support (Petracca 1992). The institutional agendas usually reflect the basic, policy, and instrument beliefs of the coalition with the most resources. However, the resources that are most useful for influencing the institutional agenda may differ from those most useful for influencing stages in the policy cycle.

Policy analysts have typically assumed that the largest and wealthiest coalitions dominate the policymaking process in a representative democracy, but influencing policymaking does not require an absolute majority. All that is needed is for one coalition to mobilize more supporters and resources than other coalitions. A few people who care intensely about a problem can outweigh many who care only a little. Sometimes a coalition can get its concerns on the institutional agenda because no one objects and it has access to a key policymaker.

Advocacy coalitions do not operate in a vacuum. Their activities are constrained by the social, legal, and political features of the society to which they belong. Since "the relationship of a particular [policy] subsystem to other subsystems and the broader policy system must be taken into account" (Sabatier 1993, 20), advocacy coalitions are often concerned about issues in several different subsystems simultaneously. Therefore, the activities of advocacy coalitions within a single policy subsystem must be considered in the light of related policy subsystems, variable political and societal factors, and the stable system context to understand how they move issues onto the agendas of government.

Policy subsystems typically contain only two to four politically significant and active advocacy coalitions, with one being dominant within the subsystem. Paul A. Sabatier (1993, 26) suggests that two advocacy coalitions are active in the U.S. clean air subsystem—the clean air coalition and the economic feasibility coalition. The clean air coalition is dominated by environmental and public health groups, their allies in Congress, most pollution control officials, and some researchers. The economic feasibility coalition, however, is dominated by industry, power companies, their allies in Congress, several labor unions, some state and local pollution control officials, and several economists (Sabatier 1993, 26). Although we generally agree with Sabatier's description of these coalitions, we have renamed each coalition to better reflect their basic concerns and their strategies in advocating their respective positions. The clean air coalition is renamed the quality of life advocacy coalition, and the economic feasibility coalition is renamed the economic development advocacy coalition. We found that these two major advocacy coalitions are active in all three policy subsystems involved in the issue of employee commute options (ECO).

These two advocacy coalitions differ in their basic beliefs about fundamental values, specifically whether economic efficiency (economic development) or social well-being (quality of life) should have top priority in public policy. They differ in other basic beliefs: production versus consumption, individual versus community, freedom versus constraints, fairness versus equality, and competition versus cooperation. They differ in their policy beliefs about what problems should be on the institutional agendas: development versus preservation, roads versus transit, and jobs versus environment. They differ in their instrument beliefs about the appropriate strategies to use in public policy: market versus government, local versus national, incentives versus penalties, and voluntary versus mandatory.

These differences do not mean that the economic development advocacy coalition does not value individual and community well-being, or that the quality of life advocacy coalition does not value economic productivity. It does mean that the two coalitions perceive and define problems and solutions differently based upon their belief systems (Sabatier 1993).

The economic development advocacy coalition comprises primarily business and industry groups, whereas the quality of life advocacy coalition includes primarily environmental and health interest groups. The economic development advocacy coalition has been most dominant in the urban and transportation subsystems since urban economies and transportation systems are high priorities of this coalition. The quality of life advocacy coalition has been most dominant in the clean air subsystem since the protection and enhancement of life, both human and nonhuman, is one of its high priorities. The issue of commute reduction involved these two advocacy coalitions in a new competition to influence public policy across the three policy subsystems. We turn now to an in-depth examination of the belief systems of the economic development and quality of life advocacy coalitions.

Economic Development Advocacy Coalition

The economic development advocacy coalition's belief system centers on fundamental values of liberty and freedom and economic values of efficiency, productivity, and prosperity. Economic productivity is an essential component of this belief system because of the basic belief that a strong economy is essential for society. Increased economic productivity and all the benefits it brings are made possible by individual and corporate property rights and freedom.

The basic beliefs of the economic development coalition influence its policy beliefs about air pollution. This coalition questions the overall seriousness of the air pollution problem and believes that the health effects of air pollution are only a problem in isolated areas and for special at-risk people. According to members of this coalition, in most places and for most people, the air is "clean enough," and those areas with special air pollution problems can be handled through technology without government intervention. This view of air pollution is reflected in the coalition's definition of the problems in the other two policy subsystems. Traffic congestion is not really much of a problem if customers, employees, and goods can move freely around and between urban areas. When they cannot, this coalition sees it as a localized problem that can be solved by adding more roads or building better traffic control systems. The quality and use of urban space relate to the needs of specific businesses and can be solved by those businesses making private decisions to adapt or move.

This coalition's instrument beliefs center on using market solutions to solve problems. The market encourages innovation, effectiveness, and efficiency in the use of personnel and materials. Technology can be harnessed to achieve goals. Businesses make decisions based on actual costs, not on moral issues of right and wrong or idealistic visions of what should be. Decisions based on concrete costs rather than abstract ideals produce the best solutions for society. According to this coalition, if air pollution is a problem, governments should determine the "proper" amount of pollution based on cost-effectiveness, cost-efficiency, and technological feasibility. Incentives, not coercion, should be the policy instruments used by governments. Public policy should be minimal, flexible, and implemented at the most local level of government possible (Sabatier 1993).

Quality of Life Advocacy Coalition

The basic beliefs of the quality of life coalition center on issues of social fairness, equality, and individual and community well-being. This advocacy coalition seeks to ensure that everyone has an equally good quality of life. The purpose of the economy and society is to help individuals live as full and as meaningful lives as possible. The protection and enhancement of life and health are top priorities for public policy. This advocacy coalition has particular interest in vulnerable populations: the elderly, children, minorities, and individuals with disabilities. This protection of vulnerable populations should be absolute and should extend beyond humans to all of nature (Sabatier 1993).

This coalition defines air pollution as a serious public health problem in many urban areas and a growing problem in rural areas. Air pollution is a problem for the general population, for susceptible individuals, and for the environment (Sabatier 1993). Members of this group believe that traffic congestion and urban sprawl are also public problems that are "out of control" and in need of immediate solutions. A key policy belief of this coalition is that problems should be stopped before they get started.

The quality of life coalition's instrument beliefs center on using controls and regulations at the highest possible level of government. The federal government is best equipped to force change since local governments do not have sufficient leverage to control the regional, national, or international businesses responsible for many public problems. This coalition has a deep distrust of business and its ability to regulate itself or change voluntarily. The coalition distrusts market solutions, believing that the market is incapable of regulating negative externalities (Sabatier 1993).

Policy Subsystems

The problems discussed on the systemic agenda and generating activity on the institutional agenda change over time. Policymakers seem to be concluding that technological solutions to public problems are limited and people's behavior will have to change. This reflects a general increase in the influence of the quality of life advocacy coalition relative to that of the economic development advocacy coalition. The progression from technological to behavioral change can be seen most clearly in the clean air subsystem but can also be traced in the transportation and urban subsystems.

Clean Air Subsystem

We consider the clean air subsystem as separate, although related to, the larger environmental policy subsystem. The latter deals with the quality of the entire natural environment and covers a wide range of subjects, from air pollution to animal extinction to nuclear waste disposal. The clean air subsystem focuses on the quality of the air, covering issues such as ozone levels, airborne toxins, particulate matter, nitrogen and sulfur oxides, and the effects of acid rain. To understand fully the dynamics within the clean air subsystem, however, we must briefly examine the larger environmental subsystem of which it is a part.

In the early days of the U.S. republic, no official policy regulated the use of the environment. People did not see any need for one. The land and its resources looked endless and inexhaustible. No one imagined a need to regulate their use. Further, nature was viewed as hostile and something to be conquered and tamed. Anything without direct utility for humans had little value (Smith 1992, 13). As the population grew and people began to see nature's resources as finite, they raised voices for restraint. They believed that too many people used common natural resources as free commodities and waste receptacles, creating a real danger of overuse (Hardin 1968). Some individuals called first for the conservation and then for the preservation of the nation's resources.

The early twentieth-century conservation movement, often called the first wave of environmentalism, grew out of the perceived need to conserve the nation's resources. It is most notably linked to John Muir (1838–1914). One of his basic beliefs was that the country should change its frontier mentality of unrestricted exploitation of nature, adopt a restrained and rational perspective regarding the use of the nation's physical resources, and carefully develop them for long-term use (Hays 1987, 13–22). Conservationists believed the cause of environmental problems was the inefficient and wasteful use of physical resources. Their main policy beliefs centered around the principles of scientific management and advanced technology. They supported such policy instruments as education, limited regulation, use permits, and large-scale technological projects such as construction of dams.[1]

The economic development advocacy coalition can be linked with environmental conservationists believing that efficient use of natural resources is important for economic profitability. The environment is an important commodity, providing raw materials for manufacturing and accepting the waste products. They also believe technology can solve any problem arising in the environment.

By the middle of the twentieth century, people outside the conservation movement became dissatisfied with the practice of "managed exploitation." They thought that too much emphasis was being placed on the use of natural resources and too little on their preservation and restoration. Their concerns were catapulted onto the systemic agenda with the publication of Rachel Carson's book *Silent Spring* in 1962. This new systemic agenda item centered on quality of life issues and marked the beginning of what has been called the second wave of environmentalism.

The roots of this second wave lay in the many social and economic changes that took place in the United States after World War II. Economic development boomed. Cities grew rapidly as the economy shifted from wartime to peacetime. People had more money to spend and more things to buy. People moved out of the dirty cities and into the clean suburbs, into newer and larger single-family homes. They vacationed in the countryside and wilderness areas. These socioeconomic changes brought more people into contact with the natural environment, developing greater sensitivity to nature and providing more public support for environmentalism.

A basic belief of environmentalists is that people are only one part of the environment, the overall "web of life," and not mere controllers of the environment. They believe that human activities place a great deal of pressure on the environment and create imbalances that can permanently damage ecosystems. Preserving the natural environment is necessary for having a good quality of life. One of their policy beliefs is that environmental problems occur because humans hold an exploitative view of nature, a view that they believe must change. Another policy belief is that human activities should be closely monitored and modified to stop the detrimental effects of human actions on natural ecosystems (Hays 1987, 27–29).

The early agenda of this second wave of environmentalists, their beliefs, focused on technological solutions to combat environmental degradation and called for the government to force business and industry to "clean up." Items on the agenda included limiting power plant emissions, requiring industrial waste byproducts to be removed from waste water before discharging the water into rivers, and setting deadlines for automobile manufacturers to produce cars with less polluting engines. Some of these technological solutions were very successful. For example, removing lead from gasoline greatly reduced the level of lead in the air.

A third wave of environmentalism began in the 1980s as behavioral solutions were offered for environmental problems. Environmentalists called for more ecologically sensitive lifestyles for everyone and promoted the idea that people must take greater "personal responsibility for the impact of daily living on the wider natural world" (Hays 1987, 30). The adoption of these new lifestyles would change the American way of living, working, and recreating. Environmentalists assert that everyday actions should be less harmful to the environment and should include such actions as recycling waste and driving less often (especially alone). This shift in emphasis from technological to behavioral solutions became a key policy belief of what we call the quality of life advocacy coalition.

This coalition, which came to dominate the clean air subsystem, includes three types of environmental interest groups. All three kinds of groups primarily want to reduce the impact of man-made pollution on the natural environment. The first type of environmental interest group, which emerged early in this century, is the "user" group, individuals who participate in various kinds of outdoor activities and want to preserve a clean environment for pursuing these activities. User groups include the Sierra Club, the Audubon Society, and the National Wildlife Federation. The second type of interest group, which emerged in the 1970s, focuses more on public education, direct intervention in legislation, and litigation. These education and litigation groups include Friends of the Environment, the League of Conservation Voters, the Natural Resources Defense Council, the Environmental Defense Fund, and the Worldwatch Institute. In the 1980s a third type of environmental interest group arose, committed to "radical actions" to protect the environment, such as spiking trees and lying down in front of bulldozers. Radical action groups include the Sea Shepherd Conservation Society, Greenpeace, and Earth First!

Early air pollution solutions focused on technology, switching fuels, and dispersal. Switching from coal to oil and natural gas produced less pollution; however, burning these fuels still produced air pollutants, sulfur and nitrogen oxides, and smoke. As more industries used oil and natural gas, less relative gain in air quality came from additional switching from coal. A second mechanism used to combat air pollution was dispersal. Taller smokestacks sent emissions higher in the air to be dispersed over a wider area. Air-polluting industries could move and release their air pollution away from the main urban population areas. In a primarily agrarian society, which still characterized the United States in the early twentieth century, dispersal of air pollution was an easy solution to the problem of urban air pollution.

Dispersal of smoke, new technologies, cleaner fuels, alternative manufacturing techniques, more efficient use of materials, and recycling waste products can be applied to stationary air pollution sources relatively easily. These stationary sources of air pollution are limited in number and have an identifiable authority. Finding out who is in charge of a particular power plant or factory and has the responsibility and authority to initiate a cleanup is easy. Mobile sources of air pollution are harder to regulate. To force producers to clean up is difficult because millions exist and they do not stay in one place.

Environmental interest groups were largely silent on the merits of commute reduction as an air pollution reduction strategy. However, the National Clean Air Coalition, organized in 1975, lobbied hard for tough clean air laws with strict enforcement and strongly influenced the successful passage of the Clean Air Act Amendments of 1977 and 1990 (CAAA-90) (Cohen 1992). Lead by a cofounder of the Natural Resources Defense Council, the National Clean Air Coalition included the Sierra Club, National Wildlife Federation, Audubon Society, United Steelworkers Union, and some church groups. Public health interest groups joined them as part of the broad quality of life advocacy coalition, which believes that air pollution has a direct negative impact on human health. The American Lung Association testified on the impact of air pollution on the incidence of lung diseases and was one of the few interest groups to explicitly support ECO and commute reduction from the very beginning of the CAAA-90 debate.

Commute reduction adds a new behavioral solution to air pollution to the clean air agenda. Members of the quality of life coalition advocate living a more ecologically responsible, and therefore a higher quality, lifestyle and want to stimulate a fundamental revolution in the way people view the world and how they should live their lives in it (Milbrath 1984, 81). They believe that people should take greater personal responsibility for the impact of their daily living on the natural world and on their own well-being (Hays 1987, 30). A key policy belief of the quality of life advocacy coalition is that the best way to deal with pollution is to prevent it. For example, driving less, carpooling, and using public transportation cuts down on auto emissions before they occur and thereby puts less pollution into the air. They pushed the idea of changing driving behavior onto the national institutional agenda as the Clean Air Act of 1970 (CAA-70) was being developed. Changing commute patterns to lessen air pollution moved onto the local institutional agenda of Pleasanton, California, in 1984 and then onto California's state institutional agenda, resulting in Regulation 15 in 1988.

After the environmental movement became well established on the U.S. political scene, the economic development coalition realized that it could no longer ignore environmental problems. Businesses became aware that their "green image" (how environmentally friendly their business practices were) could have important public relation advantages in an age of environmentalism. They also recognized that a safe and healthy working environment could protect valuable workers, boost employee morale, and consequently increase productivity. This coalition became less opposed to environmental policies and, instead, sought ways to operate environmentally—without damaging profits. Technological solutions, market controls, and the scientific management of resources promoted by the earlier conservationist movement were still the policy instruments businesses wanted to see on the institutional agenda of the clean air subsystem. They actively opposed government regulation.

The Clean Air Working Group became one of the most prominent and influential interest groups within the economic development advocacy coalition from the standpoint of commute reduction policy. Representatives of industry trade groups in Washington, D.C., created it in 1988. It actively fought each amendment and proposal for the CAAA-90 submitted by its environmental counterpart, the National Clean Air Coalition (Cohen 1992, 108–109; Switzer 1997, 116). This group testified against the ECO component of the CAAA-90. Other groups that testified against ECO during the CAAA-90 debates were the U.S. Chamber of Commerce and the National Parking Association.

Transportation Subsystem

Transportation is vitally important to the U.S. economy. It accounts for 18 percent of the total cost of some products and 8 percent of the gross national product (Coughlin 1994, 139). Since mobile sources of air pollution are a major problem in the United States, one might expect that air pollution from cars, trucks, buses, trains, and airplanes would be a major issue in the transportation subsystem. For the most part, however, this has not been the case. Historically, the primary concerns of the transportation subsystem have been those of the economic development advocacy coalition, namely, the efficient movement of people and goods and maintenance of the physical infrastructure. The main issues are interstate commerce (land, water, and air), construction and maintenance of roads and bridges, traffic management, and public transportation. Despite this broad agenda, most policy activity within this subsystem directly or indirectly involves the automobile. The national government built interstate highways. State governments construct and maintain roads, highways, and bridges. Local governments focus on the condition of city streets.

The transportation subsystem's institutional agenda includes three general types of problems: infrastructure, social, and environmental. Infrastructure problems concern the state of the physical transportation system as defined by the adequacy of roads, bridges, tracks, and airports, their condition, their construction, and their maintenance. This subsystem generally views congestion and urban sprawl as infrastructure problems. People lose time to congestion because not enough highways or streets have been built. However, it is difficult to secure right-of-ways. The ever-expanding suburbs require new roads or road extensions. The decline in use of public transportation in most cities puts a further burden on city streets as fewer and fewer people use public transportation and more and more people use the automobile to get around (Button and Rothengatter 1993).

Social problems result from the impact of transportation on the individual and are beginning to appear more often on the transportation agenda. These include accident risks, destruction or division of local communities by infrastructure construction, visual intrusion of transportation, and social inequities resulting from an auto-dominated society (Button and Rothengatter 1993).

Environmental problems involve the impact of transportation on the physical environment. Some of these effects include noise and vibration, air and water pollution, and excess depletion of natural resources, such as oil and gasoline (Button and Rothengatter 1993).

Two advocacy coalitions dominate the transportation subsystem. One leading transportation analyst calls them technocrats and behaviorists (Coughlin 1994). Another identifies them as transportation supply management (TSM), which focuses on increasing the supply of roads, and transportation demand management (TDM), which focuses on decreasing the demand for roads (Orski 1990). We believe these correspond to the two advocacy coalitions we identified in the policymaking process. Our economic development advocacy coalition corresponds to the technocrats and TSM. Our quality of life advocacy coalition is the same as the behaviorists and TDM.

The economic development advocacy coalition includes real estate interest groups, business leaders, and elected officials. The quality of life advocacy coalition includes environmental interest groups, public transit interest groups, urban planners, architects, historical preservationists, and bicycle and pedestrian interest groups. Both coalitions view transportation problems as symptoms of larger societal problems and consider transportation policy as key to promoting their particular vision of modern life. Both coalitions view traffic congestion, in particular, as providing an opportunity to redefine transportation policy and thereby address some of the larger societal problems. However, each coalition differs in its definition of transportation problems and solutions.

Members of the economic development coalition are individualists. Their basic beliefs place a high value on individual freedom and choice, arguing that as each individual does well the whole community prospers. Individual mobility is considered a "right" because it gives people the freedom to choose for themselves where they will live, work, and play. Individual mobility in an efficient and well-run transportation system is important for a healthy and growing economy.

This coalition's policy beliefs center on governmental promotion of economic development and regional growth, emphasize short-term planning for immediate problems, and favor incremental rather than comprehensive policies. They subscribe to the traditional definition of surface transportation policy, viewing it as the construction and maintenance of a physical-economic infrastructure to ease the movement of people and goods around the country.

Supply management is the economic development advocacy coalition's answer to transportation problems. Such problems are perceived as problems of inadequate supply; the solution is to increase the supply. Consequently, this coalition supports aggressive road building and public works projects at all levels of government and strongly supports technological solutions that both promote growth and mitigate its negative effects.

In contrast, members of the quality of life advocacy coalition are communitarians. They place a premium on the good of the overall community and social equity. One of their basic beliefs is that the common good must be protected from the selfishness of individuals who damage the community and the environment. They emphasize living quality over transportation efficiency, people and nature over cars, efficient use of natural resources over exploitation, and access for everyone to transportation over isolation of the disadvantaged.

This coalition's policy beliefs emphasize transportation policies that protect the environment and promote the common good. Members view surface transportation policy as a sociopolitical tool to be used to reengineer how people live and use common resources. They advocate comprehensive, long-term planning that carefully weighs the transportation benefits of policy choices against their environmental and community costs. They want transportation to serve people, not dominate them.

For members of this coalition, demand management will resolve transportation problems. They seek to limit demand by changing the lifestyles of individuals from behavior that is harmful to the environment to behavior that promotes a good quality of life for all. The quality of life advocacy coalition seeks to solve congestion by changing individual driving behavior. Some of their solutions are telecommuting, flextime, carpooling, restricted high occupancy vehicle (HOV) lanes, congestion pricing on heavily traveled roads, and the development and promotion of alternatives to the automobile such as mass transit and bicycles. Their goal is to force people to reduce their use of the roads.

For the most part, the transportation subsystem has reflected the belief system of the economic development advocacy coalition, and the focus has been on infrastructure problems and technological solutions. Infrastructure issues moved rapidly onto the transportation systemic agenda during the 1980s (Cervero 1989). The economic development coalition advocated new technological solutions to mitigate the cost of congestion, focusing on the economic costs of delays in traffic. These solutions included concrete barriers, additional lanes, better signal controls, intelligent vehicle highway systems (IVHS), radio tags and traffic sensors, traffic signals at onramps, and other methods to regulate traffic flow. However, air pollution caused by automobiles was included in neither their definition of nor their solutions to transportation problems.

The quality of life advocacy coalition became involved in the transportation subsystem as many perceived that the policies promoted by the economic development coalition were not solving transportation problems and might even be making the problems worse. They sought to get their definition of transportation problems on the governmental agenda. According to this coalition, traffic congestion is not simply a transportation problem; it is also a clean air problem. Interestingly, they use the language of the economic development coalition to promote this belief, pointing out that traffic congestion creates inefficient use of both time and fuel. Their solution, however, is behavioral rather than technological. The quality of life coalition believes that commuters should be forced to change their transportation behavior and employers forced to promote changes in their employees' behavior. The quality of life coalition succeeded in linking environmental and transportation issues at the local, state, and national levels. We will examine these linkages later in this chapter.

Urban Subsystem

The primary issues in the urban subsystem are the economy and welfare of urban residents. As air pollution began to threaten the economy, jobs, health, and safety in urban areas, clean air also became an issue in this subsystem. Urban scholars have sometimes labeled the economic development advocacy coalition as the "Progrowth Regime," and the quality of life advocacy coalition as the "Growth Management" or "Progressive Regime" (Brower, Godschalk, and Porter 1989; DeLeon 1992). Historically, the economic development coalition has dominated this subsystem, believing that the economic health of the city provides the jobs and money needed by urban residents. Since businesses can move from the urban area if the economic conditions are unfavorable to them, the government should supply them with their basic infrastructure needs (water, roads, police, and acceptable labor force) and then leave them alone. People can pursue their desired quality of life through their individual decisions of where to live and work. Further, the automobile gives them the opportunity to choose the central city, the suburbs, or even the country.

However, the quality of life advocacy coalition believes that the decisions of individuals and businesses can cause urban problems and separate people and activities into isolated homogeneous clusters. Businesses located near high-income housing are often inaccessible to low-wage workers. Demands on the transportation system are increasing faster than the supply, as are demands on the natural environment and people's time. Quality of life advocates perceive that the rational choices of individuals and businesses threaten the urban commons.

This coalition has concentrated most of its efforts at the national level. It was successful in moving urban problems of racial discrimination, housing, welfare, education, job training, law enforcement, and health care onto the national agenda during the 1960s. However, this agenda failed to solidify during the 1970s and began to disappear during the 1980s (Kaplan 1995). During the latter part of the twentieth century, many within the urban subsystem began to realize that the environmental health of the city had an impact on its economic health. Cities with severe pollution problems were unattractive to existing and future businesses and residents.

At first, this subsystem focused on stationary air pollution sources, the factories and industrial facilities that emit smoke and other pollutants. Since these are the same facilities that provide many jobs for urban residents, city officials had to balance regulating pollution with protecting jobs. Equally important to the city's economic health are the mobile sources of air pollution, the cars, truck, buses, and airplanes that move goods and people between and within cities. Mobile sources of air pollution are difficult to control because millions exist within a city. Each mobile source has an independent operator who decides when and where to drive.

Most mobile source regulations focus on the car itself and have come down from federal and state governments. Federal regulations require car manufacturers to meet strict emission standards, and state regulations require car owners to meet mobile emission tests. Local governments have focused on automobile use more as a transportation issue than as an environmental issue. Traffic regulations, adequate roads and parking, and street and bridge maintenance have been the main automobile-related issues on the urban agenda. In recent years, this has changed, and local governments have begun to address mobile source air pollution.

Another issue of vital concern in the urban subsystem is public health, the physical health of urban residents, which is greatly affected by the environmental health of the city. Air and water pollution, inadequate waste disposal, and toxic pollutants (such as radon and polychlorinated biphenyl [PCBs]) adversely affect people's physical health and well-being. Public health interest groups perceive and define these as public problems and have been quite successful in getting their perceptions, definitions, and solutions on the institutional agenda. Most of the solutions have focused on technology. Sewage treatment plants, water purification facilities, and strict waste disposal regulations brought waterborne diseases under control. Vaccines, antibiotics, and better sanitary measures brought infectious diseases under control. Since many illnesses are caused by lifestyle behaviors, such as smoking and poor diet, educating people to change these behaviors has become a new item on the agenda. Preventive health care, physical fitness, and a healthier diet are all ways for people to take greater responsibility for their own well-being (Hays 1987, 25–56).

To address the economic, environmental, and public health needs of urban areas, the urban institutional agenda has focused mainly on technological solutions. Environmental and public health regulations, better roads and mass transit, economic incentives, and more central city cultural sites and events have been used to attract and keep people and businesses in cities. However, as in the other two subsystems, political actors in the urban subsystem are looking for new ways to improve the urban environment. Behavioral solutions have joined technological solutions. Calls for behavior changes in the way people live and work in the city are being heard on both the systemic and institutional agendas. People are urged to move back into the city, recycle waste, stop smoking, eat healthier foods, use fewer pesticides in yards and gardens, and change their driving habits to reduce air pollution. The quality of life advocacy coalition has been largely responsible for getting these behavioral solutions on the urban agendas.[2]

Different Levels of Institutional Agendas

Once a public outcry about a problem has been heard and that problem is on the systemic agenda, the next step is for the problem to move onto the institutional agenda, where the outcry can be answered by governmental action. However, hundreds of institutional agendas exist within the U.S. federal system. Every level of government—national, state, and local—has its own institutional agenda, complete with its own policy problems, issues, and political actors.

The clean air institutional agendas of every level of government have, for the most part, stressed technological solutions for air pollution. One reason the new behavioral agenda is emerging is that many see problems outstripping technological solutions. A second and related factor is the activity of advocacy coalitions as they seek to get their belief systems translated into public policy.

Local Agenda

Cities bear the brunt of air pollution. Most of the sources of air pollution are found in urban areas, and atmospheric conditions tend to concentrate the pollutants in the metropolises that produce them. However, until the late nineteenth century, little public effort was directed toward reducing urban air pollution. The working poor simply endured the dirty city air, while the wealthy escaped to the country. With the coming of the automobile, more people could leave the city at night, escaping the pollution.

Some cities began doing something about air pollution in the late nineteenth century. These early clean air laws dealt with smoke and particulates and focused on technological solutions, such as prohibiting trash burning and requiring the conversion of home heating units from coal to gas (Kraft and Vig 1994). In 1881, Chicago and Cincinnati passed legislation to control smoke and soot from locomotives. Business and industry in Pittsburgh joined with local environmental interest groups to urge the city government to do something about the region's notorious air pollution problem. They realized that dirty air was bad for business as well as the environment. Public health interest groups, such as the American Lung Association, began to sound the alarm about the dangers of urban air pollution. Early in the twentieth century, Los Angeles officials began a research program to study the effects of smog. By 1950, air pollution control laws were on the institutional agendas of eighty cities (Bollens and Schmandt 1975, 151–153).

Despite these local efforts, cities have not been very successful at using technological solutions to deliver clean air to their citizens. The problem is more often one of jurisdiction and resources than a matter of will. Although some dirty air originates and remains in a particular metropolitan area, air moves and pollution moves with it. One difficulty cities have in regulating air pollution is the failure of neighboring jurisdictions to control their own air pollution. Sources of air pollution that lie nearby but outside city limits are immune to city ordinances. A city's resources may be limited even within its own jurisdiction. It may not have the technological knowledge needed to draft air pollution legislation and cannot afford hiring and training the staff needed to monitor compliance. In addition, powerful business and industry groups often oppose local air pollution legislation and threaten to leave the community if stringent laws are passed. Local public officials often resist clean air policies because they view these policies as detrimental to the community's growth and development—the primary source of jobs and greater personal income for their citizens.

State Agenda

Early state environmental agendas centered on the development and management of natural resources rather than air pollution. State departments of natural resources were given the responsibility to manage the state's water, forests, minerals, soil, fish, game, and state parks using "principles of scientific management," reflecting the influence of the conservation movement (Hays 1987). Public health concerns about pollution later expanded state environmental agendas. The new public health and air pollution control activities were initially housed in state departments of health. By the 1950s, many states had transferred pollution control activities to the departments of natural resources, which were then divided into divisions of resource management and environmental protection (Hays 1987, 441).

Again, economic interests dominated clean air policy at the state level. Industries resisted strong air pollution controls as too costly and inefficient. By the late 1960s, some states started environmental initiatives, although most state governments were more attuned to economic interests than environmental concerns. Oregon, in 1952, became the first state to enact meaningful clean air laws. California took the lead in addressing the problem of automobile exhaust and promoted a wide range of innovative environmental policies directed toward mobile air pollution sources. Other states focused on smoke and particulates. Most of these clean air policies reflected the continued reliance on technology to solve the air pollution problem. However, state interest in clean air policy remained low. By 1960, only eight states had air pollution control legislation on the books. The first priority of most state governments remained natural resources development rather than clean air (Portney 1990, 28–30; Hays 1987).

The idea of using behavior policy strategies to reduce urban air pollution appeared on the California state agenda in the late 1980s. Regulation 15, adopted in 1988 as part of California's Clean Air Act, used employer transportation controls to reduce air pollution. It required southern California employers with 100 or more workers at one site to reduce the number of cars driven alone by employees commuting to work (Green 1995, 4–5). Employers had to develop commute reduction plans and submit them for approval to the South Coast Air Quality Management District (SCAQMD), which determined whether a plan was likely to produce the required increase in the average number of persons per vehicle. SCAQMD had sole authority to formulate, enforce, and adjudicate Regulation 15, virtually telling each employer how to achieve the target ratio. Heavy fines were imposed on companies, not for failure to formulate plans or to meet target figures, but simply for submitting plans deemed inadequate by the district (Lane 1993).

Opposition to Regulation 15 was "vehement and highly vocal" (Green 1995, 7). Critics of the district argued that it micromanaged every aspect of plan development and implementation. The high costs of meeting the requirements, the inflexibility of the district, and its minute involvement in companies' commute reduction plans led critics to attack Regulation 15 as "too costly, too vague, and too arbitrary" (Lane 1993). SCAQMD was also attacked as preoccupied with the commuter-to-car ratio and ignoring the

ultimate goal of cleaning the air. Alternate, more efficient, and more eco-
nomical ways of reducing automobile pollution were not considered.

This opposition shows the need for public and private support for a
causal connection between a policy and a problem, especially when the policy seeks to change behavior. What may be most lacking in California's
experience with commute reduction policy is the public and private belief
that commute reduction would decrease air pollution.

National Agenda

Up to midcentury, air pollution was viewed as the exclusive responsibility
of state and local governments. However, the quality of life advocacy coalition was disturbed at the reluctance of most state and local governments to
adopt high air quality standards. This coalition turned to the federal government, seeking to place its policy goals on the national agenda. Their success in doing so resulted in increased involvement of the federal government in the clean air subsystem. States, however, remain important political
actors in clean air policy since all federal environmental policies rely on the
states for implementation.

The first time air pollution appeared on the federal clean air institutional agenda was in 1955 with the Air Pollution Control Act (Public Law 84-
159). Under this act, the federal government entered as an assistant to state
and local governments and provided funds for research and technical training of managerial personnel. In 1963, with passage of the original Clean
Air Act (Public Law 88-206), the federal government entered clean air policy as a regulator with the power to intervene and mediate cross-boundary
air pollution disputes between states. In 1965, the Motor Vehicle Air
Pollution Control Act (Public Law 89-272) permitted the federal government, for the first time, to establish emissions standards for new automobiles.

The idea of commute reduction, placing restrictions on driving behavior as a means to reduce air pollution, first appeared on the national institutional agenda in the Clean Air Act of 1970 and amendments of 1977. It also
appeared on a local institutional agenda in Pleasanton, California, in 1984
and on the California state institutional agenda in 1988. The specifics of
these polices are discussed in Chapter 4.

The Clean Air Act of 1970, considered one of the nation's most important pieces of clean air legislation, established lofty goals for clean air in
the United States (Melnick 1992, 91). Its primary objective was to insure
that Americans breathed cleaner air. The approach used was to set tough
emission standards for both stationary and mobile sources of air pollution
and to combine technology forcing with tight deadlines. Policymakers opti-

mistically hoped that industry could be "forced" to develop the necessary technology.

However, CAA-70 anticipated that technological solutions alone would not achieve the reductions in air pollution needed to reach the new national clean air standards. Thus, states were also required to adopt behavioral measures directed toward driving and vehicle use, namely vehicle inspection programs and transportation control measures. Vehicle inspection programs would ensure that existing cars and trucks would operate as cleanly as they were designed to be. Transportation measures would reduce the number of vehicle miles traveled each day. Vehicle inspection programs addressed the technological side of car use, whereas transportation measures addressed the behavioral side. After considerable resistance at first, vehicle inspection programs soon became common throughout the fifty states. Today, they are accepted by the public, even if reluctantly, as the cost for clean air. Yet the required transportation control measures never gained public acceptance (Yuhnke 1991, 239).

The CAA-70 transportation measures required local governments to review both individual highway projects and regional transportation plans for their impact on ambient air quality and make them consistent with the state's air quality plans. However, the federal Department of Transportation never required local agencies to conduct air quality reviews of overall transportation plans. At the initial 1971 deadline for attainment of air quality standards, none of the state air quality plans contained the required transportation control measures, and the Environmental Protection Agency (EPA) extended the deadline two years.

In 1973, the courts ordered the EPA to begin promulgating federal plans for states that had not submitted transportation control measures in their state plans (Yuhnke 1991, 242–243). Consequently, the EPA issued transportation plans for several cities, including Boston; Washington, D.C.; and Los Angeles. These plans included fuel-rationing programs and indirect source permit regulations that required federal review and approval of large mobile-source dependent facilities, such as shopping malls, stadiums, and airports. The response was swift and intense. The threat of such federal control over local development produced an immediate political backlash from states and cities. Congress reacted with a funding moratorium against EPA's implementation of parking and fuel-rationing strategies. In addition, the states challenged the EPA's transportation control measures in court. These challenges were declared moot after the 1977 Clean Air Act Amendments that significantly restricted the transportation control measures the EPA could use and effectively repealed most of the transportation initiatives undertaken by EPA to limit vehicle use. The political message to EPA was clear: "if the states failed to adopt transportation controls, federal mandates were not a politically viable remedy" (Yuhnke 1991).

It is also clear from the EPA's experience with behavioral transportation control measures in the 1970s that it is not enough for a policy to get on an institutional agenda. The policy must also be successfully implemented and enforced.

Commute Reduction: A Behavioral Solution

Policymakers have two main strategies to combat air pollution from mobile sources: changing technology and changing behavior. The economic development advocacy coalition believes that technological solutions are the best way to combat air pollution as well as easier for business and industry to implement. The quality of life advocacy coalition believes that behavioral solutions will have the greatest effect on reducing air pollution.

Technological solutions have been the main strategy on the U.S. clean air policy institutional agenda for the past twenty years at all levels of government. Battles over technical standards have overshadowed other strategies to reduce mobile air pollution. However, the quality of life advocacy coalition believes that technology, by itself, has reached the limits of its effectiveness and that new behavioral strategies must be adopted. It insists that people must change their driving behavior to reduce urban air pollution further. Transportation control measures, preferential parking for carpoolers, transit subsidies, and mandatory carpooling are all behavioral policy strategies discussed on the systemic agenda to encourage people to drive less. Yet until 1990, policymakers hesitated to impose aggressive mandatory transportation control measures to reduce the number of miles people drive their automobiles.

The CAAA-90 brought just such mandatory measures onto the national institutional agenda. Congress adopted the quality of life advocacy coalition's belief that a change in driving behavior was needed to reduce urban air pollution further. How this policy was formulated and enacted into law is the subject of the next chapter.

Conclusion

Agenda setting is the process of determining which problems receive government attention. Two major factors affect how issues move from problem perception and definition onto the systemic and institutional agendas of government and from there to the next stage of policy formulation. The first is the competition and interaction between advocacy coalitions within a policy subsystem as they seek to translate their basic, policy, and instrument beliefs into governmental action. The second factor is external to the policy subsystem and includes socioeconomic changes, social and environmental events, changes in systemwide governing alliances, and the impact of other policy subsystems (Sabatier 1993, 23). Other policy subsystems are especially important when problems overlap and bring together different policy subsystems. The interactions result in competition between beliefs about what the problems are, which issues are important, and which solutions are best. This interaction moves public response to issues from outcry to action.

ECO lies at the intersection of three policy subsystems—clean air, transportation, and urban—and reflects the complications that arise when policy subsystems are linked. Although none of the three subsystems is primarily concerned with both air pollution and driving behavior, each defines the car as part of particular problems in the policy area. In each of the policy subsystems we see a common progression from technological to behavioral solutions for these problems. The clean air subsystem initially used principles of scientific management and stringent regulations to conserve the nation's natural resources while controlling air pollution. Many now advocate more ecologically friendly behaviors. The transportation subsystem first approached problems of traffic management with more and better roads, "smart" cars, and bigger and better mass transit systems. Some within it now propose that people pay for driving certain roads at certain times, pay more for parking, carpool, and drive less. The urban subsystem has emphasized technology to deal with public health issues and economic incentives to attract business and industry to the city. Now preventive measures and healthier lifestyles are promoted to ensure better public health and a higher quality of urban life. We see advocacy coalitions in each of these subsystems that believe that technological solutions have reached their maximum effectiveness and who support behavioral solutions for public problems.

Two major advocacy coalitions are involved in all three of the policy subsystems as well as the issue of commute reduction. The economic development advocacy coalition supports technological solutions to traffic congestion and urban ozone pollution, whereas the quality of life advocacy coalition supports behavioral solutions. In 1990, the latter was successful in getting its policy beliefs onto the national institutional agenda. Its beliefs that air pollution should be linked with driving behavior translated into the ECO component of the CAAA-90.

Notes

1. Marks of the conservationists' considerable success in getting their belief system onto the systemic and institutional agendas are the departments of natural resources management that operate to this day. These departments at the national

and state levels control the use of the nation's resources. Soil conservation, refor-estation, migratory bird protection, game animal regulations, and irrigation projects are all programs supported by conservationists.

2. About the time CAAA-90 moved onto the national institutional agenda, a debate began about whether the economic development advocacy coalition's focus on local economies was hindering economic development in the various regions of the country (Ledebur and Barnes 1992; Peirce 1993; Rusk 1993). Many argued that since a global economy is now dominant in the world, regional economies are more important than city, state, or even the national economy. However, regional govern-ments do not exist to handle regional economies or the regional problems of air quality, land use, growth controls, and transportation. "The Clean Air Act and ISTEA . . . represent the first wave of federal legislation that recognizes the needs and realities of the citistate era" (Peirce 1993). The CAAA-90 recognized that air pollution was a regional issue but did not allocate funds or establish authorities to deal with it regionally and required that state governments accept the mandated responsibility. The CAAA-90 ignored the local governmental units of cities and counties. The Intermodal Surface Transportation Efficiency Act (ISTEA), in con-trast, required the designation of a metropolitan planning organization that is regional in nature as a quasi-governmental entity to receive federal ISTEA funds and to decide their distribution.

4

Policy Formulation:
From Idea to Law

Who Formulates Public Policy?

Depending on the regime type, policy can be formulated by one individual (dictatorship) or many (direct democracy). In a pluralist society like the United States that is characterized by an institutional separation of power, the actors in the policy process are more likely to be associated with particular institutions—Congress, the executive branch, and the judiciary. This focus on institutions is a necessary starting point, but it does not fully explain the dynamics of the policy formulation stage.

Initially, policy analysts assumed a kind of structural-functional institutionalism. It was easy to assume that if Congress (or any legislative body) is constitutionally assigned the responsibility for promulgating laws, then it must formulate policy. In theory, and probably in the minds of the Constitution's framers, Congress was to have been the principal policymaking body of the government. Through the nineteenth century, policymaking was generally dominated by Congress. Since the turn of the century, however, and especially in the years since the Great Depression, there has been a dramatic shift in the balance of power in the policymaking process. The twentieth century has seen a marked decline in the policymaking role of Congress in favor of the presidency, especially in the initiation of policy. Frequent references are now made to the "imperial presidency."[1]

In the past, scholars were content to analyze only the content of laws and ignore where the idea(s) came from or who would benefit, but this changed with the behavioral revolution in the social sciences in the 1950s. Analysts began to observe and document the relatively stable relationships and patterns of interaction that occurred among government agencies, interest groups, and congressional committees or subcommittees. They saw these relationships as being so durable, so impenetrable to outsiders, and so

71

autonomous that they dubbed them iron triangles, or subgovernments (see Figure 4.1).

Today, discussion of iron triangles no longer dominates most policy analysis for three main reasons: society is increasingly complex, issues cut across several policy areas, and the number of business interest groups based in Washington, D.C., has increased phenomenally. Hugh Heclo (1978) asserts that this system of subgovernments is overlaid with an amorphous system of issue networks made up of agency officials, members of Congress (and committee staffers), and interest groups as well as lawyers, consultants, academics, and technical experts. Moreover, these networks are likely to include specialists working for state and local governments, who often have more in common with their counterparts at the national level than with their own state governors or legislators. These experts are more likely to be interested in issues for intellectual or emotional reasons than monetary or financial ones.

Clean Air Policy: The Intersection of Policy Subsystems

Policy Before 1970

Until the late 1960s, all levels of government were only minimally involved in the control of environmental pollution. As we noted in Chapter 2, pollution, in general, was not perceived as a problem, partly because the environment had considerable capacity to absorb wastes without great harm to people. However, population growth, increased urbanization and industrialization, and the extensive production and use of chemicals after World War II changed this situation.

Several variable system factors contributed to an increased awareness of environmental issues. The turbulent era of the 1960s fueled political and

Figure 4.1   An Iron Triangle

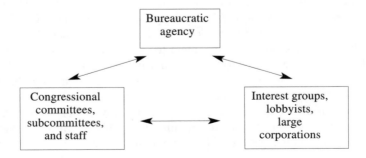

social activism, which demanded a more aggressive role for government. This activism spawned both the civil rights and environmental movements. The 1962 publication of Rachel Carson's book, *Silent Spring*, is credited with raising environmental awareness among both the general public and policymakers. The public began to associate belching smokestacks (stationary sources) and automobile tailpipes (mobile sources) with health hazards.

Until the 1960s, air pollution issues had been left up to state and local governments. The federal government's role was mostly to help fund state efforts to regulate air pollution. For example, under the Air Pollution Control Act of 1955 (Public Law 84-159), the federal government provided states with funds through the Public Health Service for research and technical training of personnel. During the 1960s, the issue of environmental pollution became a more prominent item on the national agenda.

A major expansion of federal authority occurred with the passage of the Clean Air Act of 1963 (Public Law 88-206), which continued funding for state air quality programs but added the regulatory authority to intervene in interstate air pollution problems. Two other pieces of legislation set the stage for full federal involvement in clean air policy and also established the link between the clean air, transportation, and urban subsystems. The Motor Vehicle Air Pollution Control Act of 1965 (Public Law 89-272) permitted but did not compel the U.S. Department of Health, Education, and Welfare (HEW) to set automobile emission standards. This was the first federal program to regulate automobile emissions directly (Bryner 1993). The Air Quality Act of 1967 (Public Law 90-148) established metropolitan air quality control regions (AQCRs) throughout the United States. States were authorized to establish air quality standards and develop plans to achieve them in each region. These AQCRs became the cornerstone of federal clean air policy for urban areas (Portney 1990). However, as is common in the policy process, passage of legislation does not guarantee implementation. By 1970, no state had put in place a complete set of standards for any air pollutant, and the federal government had designated less than one-third of the anticipated metropolitan air quality regions (Bryner 1993).

Policy in the 1970s

A few months after Earth Day, Congress passed the 1970 Clean Air Act (Public Law 91-604), which required draconian reductions in air pollution. The act included measures never tried before. Although most of the provisions of the act were not implemented on schedule, it set the stage for future clean air legislation. The act mandated four programs:

1. National ambient air quality standards: The Environmental Protection Agency (EPA) had to determine the acceptable ambient air con-

centration for at least seven pollutants: carbon monoxide, hydrocarbons, lead, nitrogen oxide, particulates, ozone, and sulfur oxides. These standards would be uniform across the country, and enforcement would be shared by the federal and state governments.

2. Stationary sources: EPA had to set maximum emission standards for new stationary sources on an industry-by-industry basis, issue control-technique guidelines for existing sources, and instruct the states on enforcement of the standards.

3. State implementation plans: Each state had to prepare a state implementation plan (SIP), which indicated how it would achieve the legislated goals by 1982. The 1970 act had fairly tight deadlines. It called for all areas to be in compliance with the national air quality standards by 1975, a date later extended to 1988.

4. Mobile source emissions: Automobile manufacturers had to meet detailed but flexible timetables for reducing automobile and truck tailpipe emissions. EPA and Congress granted automobile manufacturers many extensions.[2]

The 1970 Clean Air Act seemed like a "win" for the quality of life advocacy coalition. Congress had created a tough law. The automobile industry, a key player in the economic development advocacy coalition, had "lost." However, the formulation of this policy served to organize and motivate the industry. Prior to the battle over the 1970 act, automakers were in a strong financial position without competition from imports (Cohen 1992, 15–16). The automobile industry's political efforts were confined to midwestern legislators. However in 1969, General Motors established a Washington, D.C., lobbying office, and automakers began lobbying both the EPA and Congress to extend the emission deadlines, arguing that the necessary technology was simply not available.

Despite the 1970 act's emphasis on technology-forcing strategies, the quality of life coalition believed that technology alone would not be enough to clean the air. Trends in population and automobile travel projected an increase in the number of vehicle miles traveled. The coalition believed that efforts to reduce vehicle emissions involved a race between technology and behavior. Tighter emission controls reduced pollutants, but these reductions were largely offset by the large increase in the number of vehicles and vehicle miles traveled. The projections were consistent and clear: motor vehicle exhaust emissions would decline through the 1990s as a result of more vehicles meeting new standards but rise after that due to an increase in vehicle use. Thus, the coalition's policy focus began to shift from emissions standards to vehicle use. The new professional field of transportation control, or demand management (Orski 1990), became more important in the clean air subsystem as polluted cities considered how to limit the number of cars on their streets.

In 1973 the EPA published a study entitled *The Clean Air Act and Transportation Controls: An EPA White Paper.* This study reviewed the various components of the 1970 act and asserted that policymakers must look at transportation control management as a long-run complement to new federal car emission standards because "reduction in new car emission won't be enough" (USEPA 1973, 3). The study went on to point out that urban development policies that have encouraged and relied on unrestricted use of automobiles must also change. It did not delve into any of the political issues associated with any of the recommended measures or the feasibility of their implementation.

The 1977 amendments to the Clean Air Act (Public Law 95-95), the last substantive change before the 1990 amendments, did include a section, 110 (a)(3)(D), requiring urban areas to implement transportation control measures (TCM) to meet ozone and carbon monoxide standards. However, most of the nonattainment areas that included TCM in their implementation plans expected only modest yields from the programs. Moreover, there is no evidence that EPA or the states evaluated the effectiveness of the TCM programs that were implemented. According to an Office of Technology Assessment report, "It is generally difficult to evaluate the impact of individual TCMs on air quality, due to errors and uncertainty in baseline estimates of vehicle use" (U.S. Congress, OTA 1989).

The 1977 amendments also gave the auto industry two more years to meet the tailpipe standards and gave the EPA discretion to waive the stricter standards if the technology to achieve them was not available. If the 1970 Clean Air Act was a victory for the quality of life advocacy coalition, the 1977 amendments were a victory for the economic development advocacy coalition, which had succeeded in securing a discretionary loophole. Most of the strict timetables established in 1970 were extended, qualified, or reduced. The automakers, now well organized, had threatened to shut down their assembly lines rather than produce an "illegal" car (the 1978 models still did not meet the required emission standards). Urged by President Jimmy Carter to avoid potential damage to the nation's economy, Congress passed the 1977 amendments.

Cities were also unable to meet the 1970 national air quality standards. The 1977 amendments extended this deadline until 1982 and, in the case of nonattainment areas, 1987. Industrial polluters were also given three more years before they would be subject to heavy fines. Even with these extensions, it was becoming clear that some cities were simply not implementing any programs. The General Accounting Office (GAO) reported that three major cities with serious compliance problems (Charlotte, Houston, and

Los Angeles) were failing because control measures were not implemented, the control measures that were implemented were not enforced, and deficiencies identified in the three areas' ozone control programs were not corrected. The bottom line was that EPA's oversight was not as effective as it should have been. Interviews with state and local government and industry officials indicated that cities were generally reluctant to implement control measures that would involve "changes in life-styles or in an areas' industrial or business development" (USGAO 1988, 24). The concerns and beliefs of the economic development advocacy coalition had affected implementation of federal laws at the local level (USGAO 1988, 3–4).

Formulating Policy in the 1980s

The 1980s were characterized by "environmental gridlock" (Kraft 1996, 103). Ronald Reagan, a conservative Republican and former governor of California, was elected president in 1980. His election gave the economic development advocacy coalition something to celebrate. Reagan's promise to get government off the backs of the people included shifting responsibilities to the states, relying more on the private sector, and reducing the scope of government regulation. Prime targets for reevaluation were environmental regulations, which the economic development coalition believed had an adverse impact on the economy (Kraft and Vig 1994, 14).

As we noted in Chapter 1, systemic governing coalition factors affect policy formulation. In 1980, the economic development advocacy coalition gained power within the clean air subsystem when Republicans won a majority in the U.S. Senate for the first time since 1955. This enabled the economic development advocacy coalition to stop or at least delay initiatives of the more liberal and majority Democratic House of Representatives. Socioeconomic conditions also affect policy, and the recession that hit the country between 1980 and 1982 put jobs and economic growth ahead of environmental protection as major public concerns.

The Reagan administration wasted no time in reviewing environmental regulations. In March 1981, the National Commission on Air Quality, created by the 1977 Clean Air Act Amendments and filled with Reagan appointees, recommended that attainment deadlines be extended, tailpipe emission standards be lowered, and the prevention of significant deterioration program be weakened.[3] The Reagan administration's draft reauthorization proposal went even further and made enforcement lawsuits optional, eliminated the prevention of significant deterioration program, doubled allowable tailpipe emissions, and eliminated the durability requirements for motor vehicle emission control equipment. Industry representatives hailed the proposals. One industry lobbyist is reported to have said, "I don't see anything we'd object to yet" (Bryner 1993, 86).

At the agency level, Reagan appointed administrators who displayed visible contempt for the environmental policies they were charged with enforcing, according to Norman J. Vig: "Virtually all of these appointees came from the business corporations to be regulated or from legal foundations or firms that had fought environmental regulations for years" (1994, 77). Administrative discretion written into the 1970 Clean Air Act and 1977 amendments permitted industries to avoid implementation without fear of enforcement.[4] Anne Gorsuch Burford, EPA administrator, and James Watt, Secretary of the Interior, both made it clear that their agencies intended to use this discretion to help industries solve their pollution problems so that aggressive enforcement measures could be avoided.

The quality of life advocacy coalition "circled the wagons." In 1981, leaders of the nation's largest environmental organizations (called the Group of 10) met to forge an alliance against the Reagan administration and to coordinate their lobbying activities. The environmental organizations had become more professionalized since the 1970s. Membership in the groups skyrocketed with the appointments of Watt and Burford. The Sierra Club launched a "Dump Watt" campaign and collected over one million signatures in an attempt to convince Reagan to replace Watt. The Audubon Society used direct mail campaigns to attack the Reagan administration and raised ten times more funds than it had ever raised before (Switzer 1997, 186–187). Some argued that Reagan was the best thing that could ever have happened to the environmental movement.

In 1984 an accident at a chemical plant in Bhopal, India, killed 2,000 people, injured hundreds of thousands more, and focused public attention on the environment. Within three months Congress was back at work on the environment. Congressman Henry Waxman (D-Calif.) "launched an investigation, including extended hearings in which witnesses said that a Bhopal-like incident was possible in the United States" (Cohen 1992, 34). By now the EPA was considerably more open to congressional oversight. William Ruckelshaus had replaced Burford, who resigned in 1983 after being cited for contempt of Congress for her refusal to turn over subpoenaed documents.

In 1986 the worst nuclear power plant accident in history occurred at Chernobyl in the Soviet Union. Thousands of people were exposed to high levels of radiation as radioactive fallout spread across northern Europe. The accident became an important triggering event for public opinion and increased the credibility of those arguing for additional environmental protection. It provided the necessary push for environmental groups to build congressional support for stronger statutes. Although several clean air bills had been introduced during Reagan's first term, no action was taken until 1986, when Democrats regained control of the Senate, and the systemic governing coalition changed again.

In late 1987 clean air legislation reached the House floor for the first major showdown during the Reagan administration. The 1977 amendments had set 1982 as the deadline for local areas to meet the National Ambient Air Quality Standards (NAAQS) but had permitted EPA to grant an extension until 1987. The EPA could not legally extend the deadline but was reluctant to impose sanctions for failure to meet the standards, especially if local areas could show they had made a "good faith effort"—that is, the state had submitted a plan, the plan was approved by EPA, but the state failed to achieve the expected ambient air quality standard.[5] House members decided to fight over an extension of the deadline. Led by Silvio Conte (R-Mass.), the quality of life advocacy coalition wanted to extend the deadline until only September 1988, hoping to keep the pressure on Congress to tackle new clean air measures. On the other side, John P. Murtha (D-Pa.), an advocate for his home-state steel and coal companies, wanted to delay sanctions until August 1989. John Dingell (D-Mich.) lobbied actively on behalf of Murtha, but Conte's bill passed by an unexpectedly large margin (257 to 162) supported by Waxman and environmentalists (Cohen 1992, 35).

This largely symbolic vote carried important messages for both advocacy coalitions in the clean air subsystem. The votes sent a strong message to Dingell that he had less influence in the full House than he had in the Energy and Commerce Committee, which he chaired. The vote also evidenced the inadequacy of industry's lobbying efforts. Dingell is reported to have lectured many corporate representatives in Washington, D.C., that they would have to do a much better job when the inevitably broader clean air bill reached the House (Cohen 1992, 35).

The economic development advocacy coalition heard the rebuke. The Clean Air Working Group, an umbrella organization to coordinate lobbying efforts for industry groups, was formed in 1988. Under the leadership of Bill Fay, this group brought enormous resources to bear on the clean air debate in Congress. Its membership numbered 2,000 at the height of negotiations on the Clean Air Act Amendments of 1990. The group was divided into ten separate teams, each headed by a lobbyist whose industry was directly affected by the issue. Since clean air legislation threatened to impose new regulatory burdens on virtually every industry, the group tried to maintain industry cohesiveness by discouraging individual companies from cutting their own deals with environmentalists or members of Congress (Switzer 1997).

On the other side of the advocacy divide, the quality of life coalition was not idle. The National Clean Air Coalition (NCAC), founded in 1975, was well-organized under the leadership of Richard Ayres, a cofounder of the Natural Resources Defense Council. The NCAC included all the big environmental players—the Sierra Club, the Environmental Defense Fund, the National Wildlife Federation, and the Audubon Society as well as the United Steelworkers Union, the American Lung Association, and some church groups (Cohen 1992). This coalition had already battled with the Reagan administration over other environmental issues and was poised to lobby hard for a tougher clean air law and its stronger enforcement.

Thus the stage was set for a major rewrite of the 1977 Clean Air Act Amendments. Both advocacy coalitions were ready for battle. The 1988 presidential campaign was expected to have little impact on the clean air debate. George Bush, the Republican candidate and Ronald Reagan's vice president for eight years, was a former Texas oilman and not known for pro-environmental views. He was expected to continue the Reagan revolution. His Democratic counterpart was Michael Dukakis, the popular governor of Massachusetts who was lauded for his fiscal responsibility but not his environmentalism. However, things changed rather quickly.

On August 31, 1988, in a remarkable speech at Detroit Metropark near Lake Erie, Bush laid out an ambitious environmental agenda. "The time for study alone has passed," he stated, alluding to Reagan's strategy for refusing to address the problem of acid rain and the need to revise the Clean Air Act; "those who think we are powerless to do anything about the 'greenhouse' effect are forgetting about the 'White House effect'" (Vig 1994, 80–81). Bush promised that if elected president, he would act. This change in environmental views by Reagan's vice president was a formidable system factor change that shook up the clean air subsystem. No one was more surprised than the environmentalists but so was the economic development coalition, which considered Bush its ally. Bush claimed an environmental mandate to act following his election and planned to work closely with the Democratic Congress to pass new Clean Air Act amendments early in his administration (Vig 1994, 81). He met with representatives of thirty environmental organizations, who gave him a blueprint with more than seven hundred proposals for consideration. Over the next several months, Bush assembled an executive team consisting of C. Boyden Gray, his chief in-house lawyer, family friend, and "real clean air nut"; Roger Porter, his chief domestic adviser and an expert in executive branch operations; Robert Grady, his former campaign speech writer, resident expert on environmental issues, and Office of Management and Budget (OMB) executive; and William Reilly, head of the EPA and former president of the World Wildlife Fund and Conservation Foundation. The team wasted no time. They established contacts with both advocacy coalitions—the NCAC in the quality of life advocacy coalition and the Clean Air Working Group in the economic development advocacy coalition. The president's brokers brought key members of Congress into the policy formulation state—Waxman, Dingell, George Mitchell (D-Maine), Bob Dole (R-Kans.), and Max Baucus (D-Mont.). President Bush made good on his campaign promise. He and his executive team acted as policy brokers (see Chapter 1) to get both coalitions to agree to a legislative package.

On June 12, 1990, before a large audience in the East Room, President Bush unveiled the outlines of his clean air proposal. In addition to acid rain and alternative fuel proposals, he called for improved enforcement of local smog control efforts and inclusion of auto emissions control efforts. He sent the bill to Congress six weeks later. In the Senate this bill (S. 1490) was assigned to the Environment and Public Works Committee, chaired by Quentin Burdick (D-N.D.). Burdick was eighty-two years old and not that interested in the environment, so he left effective control of the clean air legislation to George Mitchell of Maine and Max Baucus of Montana. Other pro-environmental members of the committee included Joseph Lieberman of Connecticut and Frank Lautenberg of New Jersey (both Democrats) and John Chafee of Rhode Island, who was the ranking minority (Republican) member. All three came from states with well-established problems with cars, congestion, and air pollution. This Senate committee was described as a "wholly owned subsidiary of the environmental community" (Bryner 1993, 99) and wasted no time substantially rewriting the clean air measure. Two months later, Senator Baucus and twelve of his colleagues cosponsored their own version of the Clean Air Act Amendments (S. 1630).

In the House the administration's bill (H.R. 3030) was assigned to the House Energy and Commerce Committee, chaired by Dingell, the well-placed advocate for the economic development advocacy coalition. The bill, however, was sent to the subcommittee on Environmental Protection, ironically chaired by Henry Waxman, Dingell's nemesis and key member of the quality of life coalition. Waxman had already gotten a jump-start on the Clean Air Act amendments by scheduling hearings on his own version of the bill in February 1989, just a month after Bush took office. He had wanted to keep the pressure on the president and his promise to send Congress a "new" clean air act. Waxman's hearings focused on the connection between health and air pollution (U.S. Congress, House 1989a). The formulation of new clean air policy began in earnest.

Policymaking in the 1990s

President Bush signed the Clean Air Act Amendments (CAAA-90) into law on November 15, 1990. The details of this marathon event have less to do with the textbook version of how a bill becomes a law and more to do with the tenacity and perseverance of the president's policy brokers in hammering out a deal. The details of the negotiations have been chronicled both by journalists (Cohen 1992) and scholars (Bryner 1993). As Max Baucus put it, the bill was "the most comprehensive and sweeping environmental legislation that Congress has passed in this century" (Congressional Quarterly Almanac 1990, 278). President Bush took credit for his role as policy broker in what he called "the logjam that hindered progress on clean air for 13 years" (Congressional Quarterly Annual Almanac 1990, 279).

The CAAA-90 certainly is a massive statute (788 pages long), requiring more than 10,000 pages of regulations to implement. Title I alone ranks as one of the most detailed and complex laws ever enacted (Bryner 1993, 123). The CAAA-90 contains four main titles and a number of smaller titles dealing with less important issues. It substantially amended most of the provisions of the original 1970 Clean Air Act and added two new titles concerning acid precipitation (Title IV) and stratospheric ozone protection (Title VI). The four main titles of the CAAA-90 are outlined below, with employee commute options (ECO) policy included in Title II.

Title I: National Ambient Air Quality Standards. This section revised NAAQS by classifying more substances as pollutants and covering more pollution sources. The definition of major sources was changed so that smaller sources could be regulated in areas where there is serious air pollution. The designation of nonattainment areas was also revised, to six categories of ozone nonattainment, from marginal to extreme, and two each for carbon monoxide and particulate matter. Nonattainment areas were given variable but specific deadlines of three to twenty years to meet the standards. However, moderate to extreme ozone nonattainment areas were required to show specific, measurable, and steady progress toward meeting the standard. The "15 percent goal" specifically established a 15 percent reduction in emissions within the first six years and a 3 percent reduction each year thereafter. States were required to revise their state implementation plans.

Title II: Mobile sources. This section presented revised emission standards for new cars, buses, and trucks. Tailpipe emissions of hydrocarbons and nitrogen oxides were to be reduced by 35 percent and 60 percent, respectively, in some new cars by 1994 and in all cars by 1996. Title II also required the production of clean-fueled vehicles for use in fleets (fleets of ten or more vehicles that could be centrally fueled were required to use clean fuels such as methanol, ethanol, or natural gas), and the creation of pilot programs (auto manufacturers had to produce a fleet of experimental cars that would meet even more stringent emission standards for sale in southern California by 1996). By 1998 all new cars would be required to have pollution control equipment with ten-year, 100,000-mile warranties.

Title III: Hazardous air pollutants. Emission limits for all major sources of hazardous or toxic air pollutants were established in this section, which also identified 189 chemicals to be regulated by the EPA as hazardous air pollutants. The EPA was required to list the categories of industrial processes in chemical plants, oil refineries, steel plants, and other facilities that produce any of the 189 hazardous air pollutants and to issue standards for each of them by a specific deadline. Within eight years of establishing the emissions standards for industrial processes, the EPA was required to establish a second round of health-based standards for each chemical that it believed to be a carcinogen and represented a risk of at least one cancer case for every 1 million exposed individuals. Title III directed the EPA to investigate sources and effects of toxic deposition in the Great Lakes and Chesapeake Bay; required an EPA assessment of health risks from electric utility pollutants; and established programs and requirements to prevent or mitigate catastrophic chemical accidents, including the creation of the Chemical Safety Board to investigate chemical accidents and determine their causes.

Title IV: Control of acid deposition. This section established programs to limit and monitor the emission of sulfur dioxide and nitrogen oxide, both precursors to acid rain. To do so, it created a new emissions trading program for sulfur dioxide. EPA is allowed to allocate to each major coal-fired power plant an allowance for each ton of emissions permitted; sources cannot release emissions beyond the number of allowances they are given. Allowances may be traded, bought, or sold among allowance holders. The sulfur dioxide emission allowances for each of the major power plants in twenty-one states are listed in the law. Title IV also requires that the national emissions of sulfur dioxide be cut in half by the year 2000 and that the emissions of nitrogen oxides be reduced by 2 million tons per year from 1980 levels.

It is perhaps not surprising that many provisions could be buried in such a massive law. One provision buried within Title II was the ECO, or employer trip reduction program, as it was originally called. This provision (see Box 4.1) attracted little attention initially but eventually created quite a stir. The next section of this chapter focuses in detail on the formulation of this provision. Where did it come from? Whose idea was it? Why did policymakers think it would reduce levels of ozone in the air?

The ECO Provisions of the Clean Air Act Amendments of 1990

The administration's bill (H.R. 3030 and S. 1490) did not include the ECO provision, but it did include two sections on transportation control measures. Title I, Sec. 108(a)(l) stipulated:

Box 4.1   Legislation on Employee Commute Options Clean Air Act Amendments of 1990, Section 182(d)(1)(B)

Within 2 years after the enactment of the Clean Air Act Amendments of 1990 [November 15, 1990], the State shall submit a revision requiring employers in such areas [severe ozone nonattainment areas] to implement programs to reduce work-related vehicle trips and miles traveled by employees. Such revision shall be developed in accordance with guidance issued by the Administrator pursuant to section 7408(f) of this title and shall, at a minimum, require that each employer of 100 or more persons in such area increase average passenger occupancy per vehicle in commuting trips between home and workplace during peak travel periods by not less than 25 percent above the average vehicle occupancy for all such trips in the area at the time the revision is submitted. The guidance of the Administrator may specify average vehicle occupancy rates which vary for locations within nonattainment areas (suburban, center city, business district) or among areas reflecting existing occupancy rates and the availability of high occupancy modes. The revisions shall provide that each employer subject to a vehicle occupancy requirement shall submit a compliance plan within 2 years after the date the revision is submitted which will convincingly demonstrate compliance with the requirements of this paragraph not later than 4 years after such date.

The Administrator shall, from time to time, publish and make available to appropriate Federal, State, and local environmental and transportation agencies—

(A) information, prepared as appropriate after consultation with the Secretary of Transportation, regarding the emission reduction potential of transportation control measures, including but not limited to—

(i) trip-reduction ordinances;
(ii) employer-based transportation management plans;
(iii) transit improvements;
(iv) traffic-flow improvements;
(v) area-wide rideshare programs;
(vi) no-drive days;
(vii) parking-management programs;
(viii) park-and-ride and fringe parking programs;
(ix) work-schedule changes; and
(x) road-pricing and tolls;

Section 103 sec. 181 (c)(4) stated:

Within one year after any determination by the Administrator that the level of vehicle miles traveled or congestion levels in the area [nonattain-

ment area] exceed the levels projected for purposes of the area's demon-
stration of attainment or reasonable further progress (or, where applicable,
maintenance), the State shall submit a revision to the applicable implementation plan that includes a transportation demand management program for the area.

These transportation control measures were, in many respects, a continua-
tion of similar provisions found in the 1970 Clean Air Act and strengthened
by the 1977 amendments.[6] However, states had been reluctant to implement
them.

Cars and Pollution: From Technology to Behavior

In 1987, two years before intense work on the 1990 amendments began, a
respected group of public- and private-sector analysts published a report,
Commuting in America (Pisarski 1987). The study used journey-to-work
data provided by the 1980 census. It emphasized that the number of vehicle
miles traveled was increasing faster than the population and that more
working Americans preferred to drive alone to work. These findings were
corroborated by the U.S. Office of Technology Assessment (OTA) in a
report appropriately entitled, *Catching Our Breath: Next Steps for
Reducing Urban Ozone* (1989).

OTA stated that "based on current trends in population and travel . . .
the number of vehicle-miles traveled (VMT) nationwide is projected to
increase by 2 to 3 percent per year from now through 2005, resulting in a
cumulative increase of about 40 to 60 percent. Obviously, VMT growth
could have a major impact on traffic flow in urban areas, as well as on air
pollution" (U.S. Congress, OTA 1989, 177). OTA then discussed a variety
of transportation control measures, including trip reduction ordinances. It
noted the experience of Pleasanton, a medium-sized town in northern
California, that passed a trip reduction ordinance in 1984 with a goal of
reducing peak hour commuting traffic by 55 percent. The baseline assumed
everyone drove alone (clearly the most extreme case), and employers were
expected to achieve a 15 percent reduction in the first year and additional
reductions of l0 percent in each of the next four years. OTA reported that in
the second year of the program, only three companies failed to achieve the
targeted 25 percent reduction, and twelve companies had already exceeded
the fourth-year target of 45 percent (U.S. Congress, OTA 1989, 183).

These reports supported the beliefs of the quality of life advocacy
coalition that transportation control measures were a viable policy instrument for achieving the goal of cleaner air. The reports also noted the links
among the three policy subsystems—clean air, transportation, and urban.
Members of Congress who were already predisposed to expand transportation control measures cited the reports and the success story of Pleasanton

as evidence of what could be achieved with just a little regulation. They also looked at the role played by the South Coast Air Quality Management District (SCAQMD) in mandating trip reduction programs. Southern California, the symbol of automobile-oriented lifestyles for eighty years, had adopted a far-reaching experiment in 1987 aimed at reducing vehicle miles traveled and trips between home and the work site. Implementation of Regulation 15 began in 1988. The policy required all employment sites with 100 or more workers to develop and implement "trip reduction compliance plans" to increase average vehicle ridership (AVR). It required employers to change the commuting behavior of their employees by encouraging (but not requiring) employees to consider alternatives to driving to work alone (public transit, car-pooling, vanpooling, walking, telecommuting, or cycling).

The SCAQMD estimated that about 6,200 firms, agencies, and institutions employing 2.3 million workers would be subject to the regulation. Initially, evaluation studies of Regulation 15 reported a measurable impact from the initiative (Wachs and Giuliano 1992). During the first year of implementation, the AVR at 1,110 work sites increased from 1.22 to 1.25. About 69 percent of the sites experienced increases in AVR, and about 20 percent of them reported increases of more than 10 percent in their AVRs. The AVR decreased at 31 percent of the work sites. In the second year, the AVR continued to rise to 1.30 among a second smaller panel of 243 work sites.

The California regulation and evaluations gave members of Congress a model and justification for formulating transportation control measures in federal policy to deal with air pollution in about 100 urban areas that still violated ozone standards. In addition, California policy formulators and implementers had a keen interest in making sure that any mandatory trip reduction law adopted by Congress should look like California's Regulation 15. However, the question remained whether the mandatory trip reduction law would reduce air pollution significantly.

Senate Consideration of ECO

The bill (S. 1630) that emerged from the Senate Subcommittee on Environmental Protection in December of 1989 included the following ECO provision in Sec. 183(c)(4):

> Within one year after the date of classification, a revision requiring employers in such areas to implement programs to reduce work-related vehicle trips and miles traveled by employees. Such revision shall be developed in accordance with guidance issued by the Administrator pursuant to section 108(f) and shall, at a minimum, require that each employer of one hundred or more persons increase average passenger occupancy

per vehicle in commuting trips between home and workplace during peak travel periods by not less than 20 per centum above the average vehicle occupancy for all such trips in the area at the time the revision is submitted. The revision shall also provide that any employer subject to the vehicle occupancy requirement which does not achieve the required vehicle occupancy rate increase within two years after the date the revision is submitted shall be liable to the State for payment of a fee not less than $50 per month for each employee parking space provided or subsidized by each employer. Any revenue received by the State pursuant to this paragraph may only be spent to develop or implement an air pollution control program as required by section 110 and this part.

This provision was designed to produce results. Employers had to achieve the 20 percent increase in average passenger occupancy within two years or be fined. The Senate emphasized strict enforcement using monetary penalties as the key to results.

The hearings on CAAA-90 by the Senate Committee on Environment and Public Work's Subcommittee on Environmental Protection provide insight on the formulation of ECO policy. First, the opening statements made by the subcommittee chair and other members of the subcommittee show their strong pro-environmental concerns and their beliefs that more could be and should be done to reduce air pollution. Second, the large number of witnesses from California displayed a preference for that state's environmental policies, some of which were tougher than those proposed in the administration's bill. Third, subcommittee members and many witnesses focused on the need for behavior-forcing measures to complement the technology-forcing efforts that had characterized earlier clean air statutes.

Although both supporters and opponents of a bill testify at hearings, committee chairs have substantial freedom in deciding who will testify and often use hearings to build support for their beliefs about an issue (Kelman 1996). The Senate Subcommittee on Environmental Protection held four days of hearings on the CAAA-90 (September 21, 26, 27, and 28, 1989). In his opening remarks, Chairman Max Baucus showed he had read the OTA report and believed that Americans were willing to make sacrifices. He made it clear that he favored behavior-forcing options:

> We ask more of American business and the American public than the President has asked because we believe Americans are willing to do more to get rid of air pollution. (U.S. Congress, Senate 1989b, 11–12)
>
> Since 1970, the number of vehicles registered in this country has increased 58 percent, and the number of vehicle miles traveled has increased 62 percent. So cleaner cars do not necessarily mean less pollution from mobile sources. (U.S. Congress, Senate 1989d, 1)

In his remarks, Senator Joseph Lieberman (D-Conn.) indicated his belief that the automobile was at the heart of the air pollution issue and that he was predisposed to transportation controls.

Unfortunately, the emissions from cars remain, by far, the single biggest source of urban smog. The recent Office of Technology Assessment report shows that, even after application of all known controls to all sources, cars and trucks will still be the largest contributor to smog. We are becoming a nation smothered in cars: the number of cars and the number of miles which cars travel are expected to increase by 40 to 60 percent in the next 15 years. (U.S. Congress, Senate 1989d, 4).

Statements by other subcommittee members clearly reveal that they also stood with the quality of life advocacy coalition. They wanted stronger measures to ensure that states would do what was needed to clean the air. Senator Frank Lautenberg (D-N.J.) noted that his entire state was "classified as an ozone nonattainment area even though we have some of the toughest air pollution controls in the nation" (U.S. Congress, Senate 1989c, 47). Senator Chafee (R-R.I.) was concerned that the administration's bill did not go far enough. Ignoring "dry cleaners, service stations, print shops and many others—as well as automobiles ... would be in error" (U.S. Congress, Senate 1989c, 103).

A review of the hearing record reveals that eight Californians appeared out of a total of twenty-five witnesses. The number of witnesses from the state that had passed the toughest air pollution statutes was no accident. As a committee staff member noted, "they lobbied hard to be invited and tell their story" (Rechtshaffen 1994). What is evident in their comments is their strong commitment to clean air, their belief in a clear link between cars and air pollution, and their willingness to move from technology-forcing to behavior-forcing action. Pete Wilson, California's Republican senator, opened his remarks to the subcommittee by affirming the following:

> We need to amend the Act as proposed in order to mandate cooperation with the Department of Transportation to include car pooling and other alternative transportation devices, as part of an air pollution attainment plan.... Inevitably, Southern Californians will have to adopt some fundamental lifestyle changes if we are to achieve a healthy clean air environment. (U.S. Congress, Senate 1989c, 15)

Alan Cranston, California's Democratic senator, praised the efforts of SCAQMD, saying that it "has formulated an ambitious and very stringent plan to clean the air in the Los Angeles area.... In short, it will require certain quite drastic lifestyle changes for Los Angeles area residents" (U.S. Congress, Senate 1989c, 10–11).

California Attorney General John Van De Kamp summed up the current situation in written testimony:

> The State of California has a compelling interest in the Clean Air Act Amendments of 1989. We oppose provisions that might prevent California's adoption of its own standards, in some cases stricter than

> EPA's.... We recommend greater clarity and specificity in other provisions, so as to insure that EPA acts quickly and effectively to review and oversee the states' adoptions of control plans." (U.S. Congress, Senate 1989b, 419)

Van De Kamp made a strong argument for ECO provisions. His data repeated the familiar argument that technology alone would not be enough to ensure clean air. He was also concerned that the automobile emission requirements proposed in both the administration's and the Senate's bill were less stringent than those already adopted by California. Policymakers in California were afraid that the passage of a weak national law would encourage the automobile industry to delay production of "California cars" with stringent emission controls.

Not everyone who testified supported the Senate's bill, nor did everyone agree that trip reduction plans were good. Strong opposition was voiced by two organizations—the National Parking Association and the U.S. Chamber of Commerce, both part of the economic development advocacy coalition. Regina McLaurin, president of the National Parking Association, emphatically noted that "these measures ... would drain people, jobs and economic vitality out of our central cities, while ultimately doing nothing to curb overall pollution" (U.S. Congress, Senate 1989d, 332). The U.S. Chamber of Commerce pointed out that "this requirement makes the employer police and enforce how employees travel to work. The requirement will impinge on some collective bargaining agreements and will affect some small businesses; the proposed legislative language does not differentiate between 100 employees total or 100 at a given location" (U.S. Congress, Senate 1989c, 317). Both organizations emphasized the cost of implementing trip reduction programs. Clearly cost and the prospect that some small businesses might have to close their doors are always compelling arguments. Both groups also stressed the need to develop alternatives (for example, mass transit) for the solo commuter before mandating trip reduction programs.

House Consideration of ECO

In the House of Representatives, the Committee on Public Works and Transportation's Subcommittee on Investigation and Oversight held hearings on the Clean Air Act Amendments and addressed ECO, although not as extensively as the Senate. The subcommittee chair, Glenn Anderson (D-Calif.), immediately linked the clean air subsystem and the transportation subsystem in his opening statement: "I am convinced that an efficient transportation system of highways and urban mass transit ... can work hand-in-hand with state air quality implementation plans that are designed to meet our goal of clean air" (U.S. Congress, House 1989h, 3).

Since the committee chair was from California, it was not surprising that the California contingent was well represented at the hearings. Norton Younglove, chairman of the board of SCAQMD, shared his views on the state's trip reduction efforts. He filled his testimony with figures and projections of what had been and could be accomplished by encouraging commuters to double up:

> Today's commuting patterns result in an average of only 1.1 people per vehicle. Emissions go up when cars sit idling or when they operate at low speeds. On average length trips, starting a cold engine amounts to 30 percent of the pollution on that trip. Thus, it is clear that increased ridesharing is one of the methods of addressing transportation congestion issues. If we increase the average numbers of people per vehicle during rush hour from the current 1.1 to 1.5 in the Basin, we could:
>
> * reduce vehicle miles traveled by 25 percent, offsetting the projected growth in travel;
> * reduce the number of vehicle trips by nearly 34 million each day;
> * reduce emissions of reactive organic gases by up to 24 tons per day, nitrogen oxides gases by up to 34 tons per day, and carbon monoxide gases by up to 216 tons per day. (U.S. Congress, House 1989h, 263)

Written statements submitted by the Los Angeles County Transportation Commission and the Southern California Association of Governments gave the committee more evidence that changing driving behavior was an important component of the clean air equation. The commission was direct in its view that "some level of lifestyle change is necessary to improve air quality"(U.S. Congress, House 1989h, 263).

Testimony from policymakers in California seemed to confirm that Californians, at least, were committed to clean air even if that commitment dictated changes in their lifestyle. Further, the belief that a reduction in solo commuting would transfer into significant reductions in air pollution was unquestioned by most witnesses before the committee. It is perhaps easy to understand why members of Congress were persuaded that trip reduction laws made good clean air sense.

ECO After the Conference Committee

Some clean air policy observers thought the ECO provision might be dropped in the conference committee and were surprised when it emerged from conference somewhat stronger (Rechtshaffen 1994).[7] As a product of California's concerted effort to deal with persistent air pollution, ECO was viewed as an idea whose time had come. The ECO provision that emerged from the conference committee and became law was largely the Senate version. Where the Senate bill required employers to increase the average passenger occupancy per vehicle by 20 percent, the final bill raised the ante to 25 percent. However, the law gave affected states two years to submit revisions to their state implementation plans requiring employers to formulate trip reduction plans. The biggest change from the Senate bill was the elimination of sanctions, or fines, for failure to achieve the required vehicle occupancy rate. The Senate bill stipulated that employers who failed to achieve the required vehicle occupancy rates would be liable for a fee of not less than $50 per month for each employee parking space provided or subsidized by the employer. The prospect of a sizable fine was designed to give employers a significant incentive to meet the goal. The final policy formulated at the federal level contained no specific reference to a fine or penalty for not achieving the trip reduction goal. States could levy fines on employers who failed to submit plans or meet targets but were not required to do so. In other words, the teeth were removed from the policy, something that would later prove to be one of the factors that doomed mandatory ECO policy. The final bill passed the Senate by a vote of 89 to 10. It passed the House of Representatives by a vote of 401 to 25 (Cohen 1992, 166). Voting reflected strong bipartisan support.

Conclusion

Policy formulation involves the formal development of acceptable alternatives or proposed courses of action for solving public problems. In a pluralist society like the United States, policy formulation involves many people, groups, and institutions. Our policy cycle advocacy system approach, based on the centrality of policy subsystems, links the strictly institutional approach to the more amorphous system of using issue networks to understand policy formulation. Within each policy subsystem, two to four advocacy coalitions compete to shape policy. Policy formulation reflects the beliefs of the coalition that can best mobilize the resources and avoid the constraints of the larger socioeconomic system. Changes in variable system factors outside the policy subsystem can change the relative power of advocacy coalitions and shape how the policy is formulated.

We used our policy cycle advocacy system to analyze ECO policy formulation. Two competing advocacy coalitions clearly emerged within the clean air subsystem during the 1970s. Their influence can be seen throughout the 1980s and in the formulation of ECO policy in the CAAA-90. The quality of life advocacy coalition dominated the clean air subsystem during the 1970s, with successively stronger clean air provisions passing Congress. It lost dominance at the national level during the 1980s and was not always dominant at the state and local levels during the 1970s and 1980s. At the state and local levels, the economic development advocacy coalition successfully limited or delayed the implementation of many clean air policies. Two environmental accidents in the mid-1980s and the election of President Bush in 1988 allowed the quality of life advocacy coalition to regain dominance by the end of the 1980s and to formulate the CAAA-90. They successfully included the ECO provision in the broader CAAA-90. Success in policy formulation, however, is often difficult to sustain during the implementation stage. The economic development advocacy coalition did not like the CAAA-90 generally, and it particularly disliked the ECO provision. It began mobilizing resources so it could dominate and change the policy during the next stage of the policy cycle.

Notes

1. See Dodd and Oppenheimer 1985; Edwards and Wayne 1997.
2. Congress explicitly called for a 90 percent reduction in hydrocarbon and carbon monoxide emissions from the levels of 1970, to be achieved by the 1975 model year, and a 90 percent reduction in the level of nitrogen oxides by the 1976 model year. Yet it gave the EPA the authority to waive the deadlines, which the EPA did several times. See Kraft 1996, 86.
3. In 1974 the EPA issued regulations that divided metropolitan clean air regions into three categories. The air in class I areas, such as national parks, would be protected against any deterioration (commonly referred to as prevention of significant deterioration). Pollution would be permitted in class II and class III areas until national ambient air quality standards were met. This became one of the most controversial issues in clean air policy and was opposed by the Ford administration, the oil industry, electric utilities, construction interests, and other business groups. See Bryner 1993, 85.
4. The 1990 amendments that were eventually passed reflected congressional frustration with the unwillingness of EPA and many Reagan administration officials to ensure that the law was forcefully implemented.
5. This issue of "good faith efforts" played a key role in the ECO provisions because states and companies tried to argue that they "tried" but could not achieve the specified goal.
6. The 1970 Clean Air Act established a requirement for states to develop state implementation plans, which would demonstrate how they were going to attain and maintain the air quality standards, and to submit these plans to EPA for approval. The 1970 Act gave EPA the authority to set nationally uniform air quality standards and set emission standards for motor vehicles. Areas that could not demonstrate attainment by 1975 with mobile and stationary source controls alone were required to include transportation controls in the SIPs (Public Law 91-604 1970, sec. 110 (a)(2)(B).
7. Bills passed by both houses of Congress often differ. Such differences must be resolved in a joint House-Senate conference committee before a bill can be sent

to the president. Conference committees consist of representatives from both the House and the Senate, from the committees that dealt with the bill. Since conference committee members are usually chosen from among supporters of the bill, a strong incentive for compromise exists. Each chamber then votes the resultant conference report up or down (no amendments are allowed).

5

Policy Implementation:
The Public Actors

Policy implementation is neither a routine nor a highly predictable stage. It is the effort to put a policy into practice. Policy implementation can be defined as "what happens after a bill becomes a law" (Anderson 1997, 214). The time frame for implementation varies depending on the specific policy. The Clean Air Act Amendments of 1990 (CAAA-90) established numerous timetables for achieving goals. Governors were given 120 days to submit a list of ozone attainment and nonattainment areas within their state for the shortest timetable. The longest timetable for nonattainment areas was twenty years (until the year 2010) to meet clean air standards (Public Law 101-549, November 15, 1990, sec. 102).

Efforts to understand policy implementation have been hindered by the low visibility and limited accessibility of federal, state, and local government agencies assigned the task. Much that occurs during policy implementation seems tedious or mundane and may be done with limited public awareness. Nonetheless, the consequences of implementation for the substance of policy are every bit as important as what happens in the formulation stage. Moreover, closer examination reveals that strong and sometimes bitter political struggles attend the implementation of particular policies. Groups that suffer losses during formulation may seek to recoup some of those losses during the implementation stage. Automobile manufacturers delayed for decades the implementation of the strict national emissions standards for mobile sources established in the 1970 Clean Air Act by appealing to the Environmental Protection Agency (EPA) to waive the deadlines. In 1990 a presidential Council on Competitiveness headed by Vice President Dan Quayle regularly reviewed agency rules and regulations to reduce their regulatory burden on businesses. According to Michael Kraft (1996, 115), the council urged the EPA to make more than 100 changes to EPA air pollution regulations sought by the business community, including some that Congress explicitly rejected in approving the act.

In this chapter we focus on the public actors involved in the implementation of the employee commute options (ECO) program. (Chapter 6 focuses on the private actors—the companies and individuals whose behavior was the target of ECO.) Public agencies at two different levels of government translated and interpreted ECO policy as formulated in the CAAA-90. At the federal level, the EPA was assigned the task of issuing guidance to the affected state governments. State governments were then charged with converting the EPA guidance (i.e., its interpretation of the law) into regulations that would require that public and private companies and organizations develop plans to change the commuting behavior of their employees. ECO was unique because it made employers responsible for changing this behavior.

This chapter begins with a discussion of the implementation framework that seeks to answer two questions: What are the preconditions for successful policy implementation? Who is involved in the process? At this stage of the policy cycle, advocacy coalitions play important roles as they seek to gain or maintain their influence. In the second section, we examine the intergovernmental dimension of policy implementation in the United States by asking, How does federalism affect the policy implementation stage? The third section analyzes agency rule making, specifically EPA's role in developing ECO guidance. The fourth section examines how the regulations developed by four of the eleven states required to implement ECO legislation helped or hindered program implementation by the businesses and individuals.

The Implementation Framework

U.S. government is a maze of overlapping and interconnecting policymaking structures. Executive, legislative, judicial, regulatory, and bureaucratic entities compete for and share authority horizontally at each level of government. National, state, and local governments must also interact vertically as the different levels compete for and share authority. Implementation efforts move horizontally within levels of government and vertically between levels of government. The requirements of federalism and the multifaceted web of intergovernmental relations make it likely that most public policies will be implemented by a series of institutions and levels of government (Walker 1995). Multiple government institutions and agencies provide myriad points of pressure and influence for advocacy coalitions to challenge the implementation of policy.

Policy implementation is neither routine nor predicable, and no general theory of implementation has been accepted by most scholars in the field (Nakamura and Smallwood 1980; Goggins 1987). The field has benefited, however, from various efforts to develop methodological tools to investigate the implementation stage (Van Meter and Horn 1975; Elmore 1979; Yin 1985; Ingram and Schneider 1988). In this section we discuss three approaches to implementation research—the top-down model, the bottom-up model, and the advocacy coalition framework (ACF) as a synthesis between the first two.

Top-Down Approach

The top-down approach, developed by Daniel Mazmanian and Paul Sabatier (1981, 1983; Sabatier 1986), provides a useful checklist for the implementation stage. This approach identifies legal, political, and tractability (manageability of the problem) variables that affect implementation. It then distills them to a shorter list of six sufficient and generally necessary conditions (Mazmanian and Sabatier 1983, 3–24) for effective implementation of legal objectives. These six conditions follow:

Clear and consistent objectives. Laws, rules, and regulations that have precise objectives (what is to be accomplished) and that are clearly ranked in importance serve as unambiguous directives to implementing officials. Communication is an important component of this condition (Edwards 1980). Orders to implement policies must be clear, accurate, and consistent. Confusion by implementers about what to do increases the chances that they will not implement the policy as intended. Subsequent studies have shown, however, that very few policies meet this criterion (Sabatier 1986).

Adequate causal theory. Every policy incorporates, at least implicitly, a causal theory of what created the problem and what actions will reduce the problem. An adequate causal theory first requires that the links between the problem and solution be established. Second, it requires that officials responsible for implementing the policy have the jurisdiction necessary to attain the objective. When policy requires technology that is not available, officials lose control of the causal relationship, and implementation is delayed.

Centralization of control and resources. Implementation is enhanced when the policy charges a single unit with carrying out the policy and gives that unit ample inducement and the ability to sanction those flouting the policy.[1] When many government units must be involved, the policy can make one of them legally responsible, with the other agencies or subdivisions reporting to the one. Legally mandated inducements and sanctions increase the likelihood of program implementation. Putting incentives (monetary) and sanctions (fines) into the legislation ensures that those assigned the task of implementation will not waver (much) from the original goals of the program.

Committed and capable implementing officials. Probably the most important resource in implementing a policy is skillful staff. Policy implementation in both the private and the public sector can be hampered by inadequate staff. Staff must possess the background necessary for the job (education, expertise) and the disposition (willingness) to administer the policy and the necessary resources (time, equipment, funds). Sometimes government finds it difficult to attract and retain adequate staff when government itself is under attack and there is no prestige in working for it.

Support of interest groups and sovereigns. Implementation success depends on the political support of interest groups and legislative and executive sovereigns. Maintaining this support is crucial. A major difficulty in implementing policy that involves different levels of government is that the different implementing agencies are responsible to different sovereigns, and these different sovereigns may have different interests. The head of a state agency implementing a federal policy must serve two masters, the governor of the state and the president of the United States.

Changes in socioeconomic conditions. Although abrupt political changes are not commonplace in U.S. politics, new political, economic, or social events can affect implementation of policy. The election of a conservative Republican Congress in 1994 was such an event. New political leadership interpreted the election as public support for the "Contract with America," which had been signed by then–Minority Leader Newt Gingrich and 300 Republican candidates for the House of Representatives. The contract outlined specific policy areas in which Republicans would make changes if elected. One area was unfunded mandates, such as ECO, which Republicans promised to impose on the states less often.

Bottom-Up Approach

The bottom-up approach to policy implementation focuses on the point at which administrative actions intersect private choices (Elmore 1982, 21). It begins not with the statement of intent coming out of policy formulation but with the specific behavior occuring at the lowest level of policy implementation. This approach is based on the premise that the interaction of lower-level officials with their clients or target populations is the key to understanding policy implementation. Lower-level officials, sometimes called "street-level bureaucrats" (Lipsky 1980) must enforce a law for implementation to occur. Enforcing the law, however, may not cause change or even be the cause of observed change. Adherents of the bottom-up approach assume that policy is only one influence, and perhaps only a minor one, on the behavior of people engaged in the implementation process. They contend that state and local economic conditions, the attitudes of local officials, and the actions of clients have greater effects on implementation than the policy itself (Hjern and Hull 1982; Elmore, Elmore 1987). The bottom-up approach (Hjern and Hull 1982) begins by

- identifying the network of local actors associated with a particular policy area;
- surveying these actors about their goals, strategies, activities, and contacts; and
- generating an implementation structure of local, regional, and national actors.

The bottom-up approach works better than the top-down approach for assessing the dynamics of local variation since it focuses on local implementation structure. However, the extensive interviewing of participants required by this approach consumes time and is expensive.

Since neither of these models alone satisfactorily explains policy implementation, several efforts have emerged to combine the two approaches (Elmore 1982, 1987; Sabatier 1986). Richard Elmore argues that policymakers need to consider both the policy instruments and other resources at their disposal (top-down) and the incentive structure of ultimate target groups (bottom-up) because policy success is contingent on meshing the two.

The Advocacy Coalition Framework

Our model also integrates the two approaches, following the advocacy coalition framework discussed in Chapter 1 as a major contribution to our general approach to policy analysis and applying it here to policy implementation. ACF proposes that policy subsystems are the most useful unit of analysis for understanding public policy since they include all the actors involved—advocacy coalitions, agencies at all levels of government, private businesses, and individual people. National public actors provide only a part of the implementation picture. Street-level bureaucrats provide only part. Most national domestic programs rely heavily on subnational governments for actual implementation, and these governments are substantially represented in groups lobbying national legislatures and agencies. Private-sector firms and individuals are involved in carrying out policy and also in lobbying the different levels of government. By including these actors in our analysis, we can achieve a more accurate picture of policy implementation.

We began our evaluation, however, with only a bottom-up approach and identified a random sample of companies in one affected region to study how they implemented policy. Two of the authors of this book were on the original evaluation team that believed the real test of ECO policy implementation was whether employees changed their commuting behavior. The team soon realized the importance of other parts of the policy subsystem and asked the third author to join the evaluation and bring analytical tools appropriate for a comprehensive analysis. We learned that the roles of federal, state, and local governments in the implementation of national policy are as important as those of businesses and individuals. Therefore, in this chapter we utilize a top-down analysis of key decisions by government agencies that affected the implementation of ECO.

The Intergovernmental Dimension

In *The Federalist Papers,* no. 17, Alexander Hamilton noted that "it will always be far more easy for the State government to encroach upon the national authorities than for the national government to encroach upon the State authorities" (O'Connor and Sabato 1995, 90). Few people, and probably no state administrators, would agree with this statement today. Through a variety of mechanisms—grants-in-aid, preemption, and mandates—the national government has encroached onto policy areas formerly considered the domain of state or local governments or outside the realm of government. This growth in federal involvement has expanded the number, size, and scope of policy subsystems and has also led to the emergence of well-defined advocacy coalitions.

Although most people think of Washington, D.C., when they hear the word *government,* state and local governments provide most of the services that people use (Dye 1990; Kincaid 1990; Walker 1995). Many policies are formulated by national officials, but they are implemented by state and local governments that exercise discretion over the precise forms the policies take (Boeckelman 1992; Anton 1989). For example, the 1970 Clean Air Act allowed each state to develop its own plan for the implementation and enforcement of standards within each of the air quality control regions. The use of air quality standards instead of emission limits gave the states an opportunity to choose any pollution control mix (mobile or stationary) adequate to meet the standards. One consequence of this state-based system is wide interstate variation in many aspects of environmental policy. Here lay both the promise and peril of federalism.

The promise of federalism is the opportunity for diversity. We often characterize state governments as "laboratories of experimentation" (Osborne 1988), where prospective federal policies can be tried out on a smaller scale and existing federal programs can be adapted to the conditions and needs of individual states. States are closer to the people than the federal government and can tailor policies to meet people's needs. Diversity at the state level can be valuable in encouraging innovation.

The peril of federalism is that states differ in the resources they can devote or are willing to devote to particular problem areas. The expenditures by one state to reduce a problem may produce few benefits if neighboring states do not make similar expenditures. Many of our most difficult air quality, transportation, and urban problems overlap state boundaries, and attempts to deal with these problems comprehensively requires some uniformity of effort. Without federally mandated uniform policy, states and local governments have few incentives to work together and coordinate policy, especially if the policy involves restrictive regulations (Anton 1989). Since states compete with one another for business, any state with a less restrictive regulatory environment vis-à-vis other states places itself at a competitive advantage. Businesses can and do travel across state lines to improve their bottom line (Bloksberg 1989; Dodson and Mueller 1993). State and local governments are often willing to reduce the regulatory and tax burden on businesses they seek to attract or retain, even at the risk of sacrificing other public goals. Diversity can be detrimental in a problem area by encouraging a state to do less than its neighbors to solve the problem. This is what Harvard professor Paul E. Peterson calls "a race to the bottom" (1995).

A national program with uniform requirements provides no incentive for businesses to move to another state. Some multistate businesses even prefer uniform regulatory standards since they permit businesses to adopt one set of operating procedures.

Congress gave the EPA the responsibility to translate the ECO policy into guidance for the states. How much discretion should EPA give to state implementing agencies? What did states do with the required degree of uniformity and the flexibility they were given? What role did advocacy coalitions play in the development of federal and state agency guidelines and regulations? The next section begins with a brief overview of the federal regulatory process and then reviews how the EPA developed ECO guidelines.

Policy Translation: From Law to Guidance

Implementing laws, executive orders, and judicial decisions requires agency officials to interpret often vague language and develop the means to achieve the intended goals. Typically this involves drafting rules that are legally binding. The 788 pages of the CAAA-90 required 10,000 additional pages of regulations before being fully implemented (Rosenbaum 1995, 14). Every major provision of the act, including ECO, required subsequent rules to clarify and refine it.

Rule Making

When Congress creates a department or agency, it delegates some of its powers listed in Article I, Section 8, of the U.S. Constitution. Congress recognizes that it does not have the time, expertise, or ability to involve itself in every detail of every program; therefore, it sets general guidelines for agency action and leaves it to the agency to work out the details. Thus, most implementation involves administrative discretion, the ability of agencies to make choices concerning the best way to implement congressional intentions. This discretion exists even in legislation as detailed as the CAAA-90.

Administrative agencies exercise their authority through formal rule making, the administrative process that results in regulations. Rule making is a quasi-legislative process. Regulations govern the operation of all government programs and have the force of law. In essence, administrative rule makers act as lawmakers when they draft regulations to implement congressional statutes. Since regulations often involve political conflict, the 1946 Administrative Procedures Act established federal rule-making procedures to give everyone the chance to participate in the process. Similar procedures are used by state and local agencies. The act requires that

- public notices of time, place, and nature of rule making be published in the *Federal Register*;
- interested parties be given the opportunity to submit written arguments and facts;
- the statutory purpose and basis of the rule be stated; and
- the final rule be published at least thirty days before it takes effect.

Although these requirements are designed to allow everyone and anyone to participate, the reality is that participation is usually dominated by the actors with the most knowledge and interest in the issue (Kerwin 1994). Most people find it difficult to understand the technical nature of the issues subject to rule making, so experts dominate.

Administrative agencies also exercise formal authority through administrative adjudication. This is a quasi-judicial process in which government agencies settle disputes between two parties similar to the way courts resolve disputes. It cannot be a true judicial process, however, because that would violate the constitutional principle of separation of powers.

These formal administrative processes for due process and protection of individual rights are complicated. Subsystem actors frequently seek informal processes to carry out and influence policy. Administrative agencies exercise authority informally through the practical day-to-day decisions made by individual government employees. Lawyers decide whether to prosecute. Officials decide when to file noncompliance charges. Investigators check whether a site is in compliance. Officials interpret a law or a regulation. Informal discussions between agency staff and regulated parties (individuals, businesses, or state and local governments) are common. These informal contacts, letters, memorandums, and telephone calls may occur regularly when affected parties are trying to understand what they need to do to meet the requirements of a new regulation.

The larger and better-financed trade associations, industries, and other organizations employ an army of Washington, D.C., law firms and technical consultants to help make their case through both formal and informal administrative procedures. Although environmental and citizen groups are rarely as well financed as business interests, they often have significant opportunities to shape the outcome, especially when decisionmakers believe that these groups speak for the public.

ECO Guidance

The EPA had one paragraph in the CAAA-90 from which to develop the requirements of the ECO program and guide affected states in developing state regulations. (Refer to Box 4.1 for the ECO law.) The affected states were required to promulgate ECO regulations or rules and to submit them to EPA for approval as part of the state implementation plan (SIP). States had to "convincingly demonstrate" (USEPA 1992, 13) how they planned to achieve the ECO requirements of the Clean Air Act. If EPA was not convinced that the state could achieve its clean air goals, the SIP would not be approved, the state would be violating the CAAA-90, and the state would risk losing federal highway construction funds. Thus states had a definite incentive to take EPA's guidance seriously.

EPA based its decision to issue ECO "guidance" rather than the customary "regulations" on the specific statutory language found in the law. Unlike other sections of the act, in which the Congress specifically states, "The Administrator shall promulgate regulations..." (Subpart 5: 104Stat.2462), the ECO paragraph specifically states, "Such revision shall be developed in accordance with *guidance* issued by the Administrator...." Later in the same paragraph, the law again mentions that the "*guidance* of the Administrator may specify..." This leaves little doubt that Congress intended the EPA administrator to develop guidelines rather than more rigorous regulations.

Although it has been suggested that some agencies try to evade the rule-making function by resorting to guidelines, interpretations, or technical manuals (Kerwin 1994, 74), this does not appear to have been the case with ECO. The tight implementation/compliance schedule specified in the act also attests that EPA would not have had time for formal rule making.[2] The six-year period did not give companies much time to increase vehicle occupancy by 25 percent, even under the most optimistic scenario. It definitely would not be much time if EPA had to develop regulations and hold formal hearings.

The EPA assigned the task of guidance preparation to its Office of Air and Radiation in Washington, D.C., which in turn assigned it to its Office of Mobile Sources in Ann Arbor, Michigan. The Office of Mobile Sources took the lead in developing the guidance and involved the Office of Air Quality Planning and Standards at Research Triangle Park, North Carolina, in the process. Affected states, however, communicated mostly with their EPA regional offices, whose staff frequently interpreted the guidance differently from headquarters' staff.

EPA staff began the development process by reviewing the public record, including congressional hearings, testimony, publications, and studies considered by Congress. This is the typical first step employed by agencies as they determine the "intent of Congress." Soon EPA needed more. Stephen Breyer, a noted jurist and current Supreme Court justice, identifies four generic sources of information available to rule makers: industry, in-house research offices, outside experts, and public interest groups (Kerwin 1994, 102). EPA employed all four sources, although not equally, to develop the ECO guidance. Congressional deliberations frequently mentioned two studies: the Office of Technology Assessment's Report, *Catching Our Breath: Next Steps for Reducing Urban Ozone* (U.S. Congress, OTA 1989); and the Eno Transportation Foundation's publication, *Commuting in America* (Pisarski 1987). EPA used these two studies as well as a 1990 study by the Federal Highway Administration evaluating travel demand management measures to relieve congestion (COMSIS 1990) and two studies contracted by the U.S. Department of Transportation (DOT) on transportation demand management (Cambridge Systematic 1991).

California's South Coast Air Quality Management District (SCAQMD) was another key source of information. California was already involved in an ECO program that had many components included in the federal statute—employers with more than 100 persons, average vehicle occupancy, peak travel times, and the like. EPA staff initiated extensive communication with SCAQMD staff in developing the guidance. EPA then hired staff from SCAQMD as independent consultants to present four two-day workshops for interested "stakeholders" (companies, state officials, and transportation management associations) on ECO implementation.

EPA established a formal relationship with the National Association of Regional Councils (NARC), which was asked to review the draft guidance and comment. NARC was subsequently given a grant from the EPA to develop a series of workshops on ECO implementation and to publish a joint EPA/DOT newsletter—*The Clean Air Transportation Report.* EPA also consulted with the Association of Commuter Transportation (ACT), COMSIS, and Cambridge Systematic.

What the Guidance Said

Government agencies have discretion to interpret the laws they are charged with implementing, unlike courts that must rely on a literal reading of the law. The final ECO guidance makes clear that it is EPA's interpretation of the law. The guidance begins with a disclaimer:

> This guidance is intended to assist States in developing approvable SIP revisions but does not establish or affect legal rights or obligations. It does not establish a binding norm and it is not finally determinative of the issues addressed. EPA approval of any particular SIP revision will be made by applying the applicable law to the specific provision in the SIP. (USEPA 1992, 3)

It goes on to explain what states had to do. The guidance included dates and times when state actions were due and defined terms to make sure that everyone knew what they meant. Precise terms are important in implementation. If the same term can be interpreted differently by each state, then the prospects of achieving the desired results decrease. Unclear terms can lead to inequities as states find it easier or harder to meet the requirements depending on the definitions they use.

Definition of Terms

The details of implementing public policies often became overwhelming. ECO was much more complicated than a simple carpooling effort. Two terms that appeared in the ECO portion of the CAAA-90 illustrate how agency interpretations of the details affect the implementation of policy as law is translated into practice.

Peak travel periods. Most people will have a general idea of what Congress intended when it used this term in the law. Implementation, however, requires a specific definition rather than just a general idea. EPA added the precision not contained in the law but narrowed the scope in the process by removing the commute home from the definition of peak travel periods. EPA's definition was those hours between which the morning commute occurs Monday through Friday. EPA believed that the intent of the Act is to significantly reduce single occupancy vehicle commute trips to and from work and has defined the peak travel period to include either the hours between 6:00 A.M. and 10:00 A.M. or any other period which captures 85% of commute trips between 5:00 A.M. and 11:00 A.M. as determined by the state. (USEPA 1992, 6)

Employer. The law refers to "each employer of 100 or more persons." According to EPA guidance,

there is no indication that Congress meant the 100-employee criterion to be applied rigidly in situations in which the majority of an employer's workforce follows a nonstandard schedule.... An employer of 100 workers split evenly between three shifts would have about 33 employees arriving during the peak period. It is *EPA's judgment* [emphasis added] that fewer than 33 employees who report to work during the peak travel period do not constitute enough employees at that time for an employer to implement a viable trip reduction program and that such a situation is de minimis. (USEPA 1992, 7–9)

Thus EPA set thirty-three as the minimum number of employees who must report to work between 5 and 11 A.M. for a viable program. Shifting the schedule of a couple of employees could exempt companies otherwise subject to ECO.

The Formula for Compliance

Most of the EPA guidance was designed to clarify ECO requirements by giving states a series of "easy" steps for determining their compliance formula. The law said that the affected areas had to "increase their average passenger occupancy [APO] per vehicle... by not less than 25 percent above the average vehicle occupancy." How would the EPA know when the area has increased its APO by at least 25 percent? EPA outlined a formula that included more definitions and divided the process into five basic steps:

1. Determine the area average vehicle occupancy (AVO)—the average number of employees in the whole region divided by the number of private vehicles they use to get to work.
2. Determine the target APO—the baseline goal set for each employer, determined by multiplying the AVO by 1.25 (25 percent increase).
3. Determine the measured APO—the number of employees reporting to the employer's work site during the peak travel period divided by the number of private vehicles the employees use to get to work.

4. Calculate vehicle mile credits—the amount by which the work site meets, exceeds, or fails to reach the target APO.[3]
5. Adjust the measured APO by any credits such that:

$$APO = \frac{\text{\# of employees reporting to work during peak hours}}{\text{\# of vehicles in which employees report +/- APO credits}}$$

To the casual reader this "simple" formula may seem complex. In fact, even more informed readers (e.g., state implementation staff) found the formula and terminology initially confusing. As information and training in ECO increased, state staff became more adept at calculating average vehicle occupancy, average passenger occupancy, and so on.

EPA Suggestions for Determining Employer Compliance

EPA suggested four options that affected states could use to "convincingly demonstrate" that compliance would be met. Although the guidance did not require states to use any of them, the state would have to convince EPA that any other option would achieve the goals in the law. Otherwise, EPA would not approve the SIP. The four options were as follows:

- *Plan-by-Plan Review:* The state would be required to review each employer's ECO plan to make sure that the proposed methods (incentives and disincentives for employees) were achievable.
- *Performance Standards:* The state would set a performance standard for each employer to achieve. Penalties would be imposed if the employer failed to reach the standard.
- *Minimum Measures:* Under this option, the state would require all employers to implement a minimum set of measures (e.g., carpool locator programs, preferred carpool parking, increase cost of paid parking). Employers could implement additional measures, but all would have to do something.
- *Contingency Plan:* This option would be reserved for those employers who failed to meet their specific target APO. It would identify financial penalties and compliance incentives sufficiently large to convince employers to increase their efforts.

EPA expected state regulations to include penalties for an employer who failed to submit a plan or who failed to implement a plan: "Penalties should be severe enough to provide an adequate incentive for employers to comply and no less than the expected cost of compliance" (USEPA 1992, 15). The guidance was specific and made it clear that EPA would not approve just any SIP submitted by the states. EPA's power of interpretation is evidenced in correspondence between EPA's director of its Air, Radiation and Toxics Division and the Delaware Transportation Authority, dated May 1993. The EPA official writes:

> EPA's December 1992 ECO Guidance states that financial penalties in ECO programs must be "severe enough to provide an adequate incentive for employers to comply and no less than the expected cost of compliance." Under Section 1905 (c) of Delaware's draft legislation, non-complying employers would not be subject to penalty liability for the entire period that they are in violation of ECO requirements up until 30 days after the notice of the violation in question.... We are concerned that the proposed grace period could undercut the compliance incentive provided by the penalties, as it might lead some employers to believe that it is in their economic interest to delay developing or implementing their ECO plan until after they receive a notice of violation. (USEPA 1993d)

EPA initially took a very firm stand on compliance. In another informal review, September 1993, the EPA Region 5 administrator sent a letter to the secretary of the Illinois Department of Transportation:

> USEPA takes seriously its responsibility to ensure that the requirements of the Clean Air Act are met in a timely manner. Because serious approvability issues still exist, and because Illinois does not appear to be making expeditious progress in implementing its ECO program, I must inform you that if the Illinois Legislature fails to address the identified approvability issues this year, or if Illinois submits final regulations that are not approvable, then USEPA intends to use its discretionary authority under section 110 (m) of the Clean Air Act to propose the imposition of sanctions. (USEPA 1993e)

By June 1994, EPA's position on ECO was more conciliatory. A letter to Senator Frank Lautenberg of Connecticut from EPA administrator Carol Browner begins:

> I welcome the opportunity to respond to your comments about flexibility in program implementation.... Our continuing effort here at EPA is to make the program work in ways that make sense at the local level.
> States and businesses should be provided with the maximum flexibility available under the law in implementing this program (USEPA 1994).

Throughout the remainder of 1994 and during 1995, especially after the inauguration of a Republican Congress, EPA tried to convince ECO stakeholders (members of the clean air, transportation, and urban subsystems) that it was serious about flexibility. In March 1995, the EPA asked the Clean Air Act Advisory Committee to form a working group to explore ways the EPA could give states additional ECO flexibility. It included individuals representing business and industry, state governments, local governments, public interest groups, and academics. EPA adopted all five recommendations proposed by the advisory committee. Two of the recommendations seemed to change the intent of the legislation (CAAAC 1995). The first allowed states and employers to substitute equivalent emissions reductions instead of submitting a plan to increase vehicle occupancy. If a state could reduce ozone emissions from another source (such as a factory), then it would not have to implement an ECO program. Similarly, an employer in a state with ECO requirements would not have to implement an ECO program if it could reduce emissions another way. The second allowed states to claim credit for reductions in trips unrelated to work (USEPA 1995b), even though the ECO law specifically mentioned "work-related vehicle trips."

EPA's discretion to interpret the law was not limited to the initial issuance of ECO guidance but continued during the state implementation phase. As dissatisfaction, criticism, and confusion about ECO increased, EPA was more willing to reinterpret its own guidance and allow more flexibility to the states.

Disagreements Within Policy Subsystems: EPA vs. EPA

Agency discretion may lead to different interpretations of the same term within the same agency. In a memo dated December 6, 1993, the EPA Region 2 director of the Air and Waste Management Division in Boston wrote to the director of the Office of Mobile Sources in Ann Arbor, Michigan, disagreeing vehemently with the latter's ECO interpretation. The basis for the memo was a disagreement over the interpretation of the term *area*. Could a company trade ECO credits with a company in a different nonattainment area? One said "no"; the other said "yes." Two paragraphs of the lengthy memo illustrate this difference:

> I am writing to ask that you reconsider the recent interpretations made by your Office with regard to Employee Commute Option (ECO) program implementation issues.... [ECO] has very high symbolic value in terms of sensitizing the public (the employees, the employers, their families and their friends) to the relationship between the use of private vehicles and air quality. This is one of the reasons why unnecessary and avoidable negative feedback from implementation of the program in a burdensome way should be avoided. I believe that EPA should be giving the states the widest latitude possible in designing their ECO programs....
>
> By very tightly constraining how the states can implement the ECO program we change the message from one of environmental protection to one of federal intervention. The focus will be on the few "horror stories" where EPA's interpretation of the law leads to a ridiculous conclusion and not on what we (and presumably Congress) are trying to achieve. With a rigid reading of the law, we could very easily temporarily gain a very tiny extra emission reduction credit and lose the whole program. (USEPA 1993f)

More than a disagreement over terminology, this was a disagreement over strategy. One EPA office seemed to be saying that to save the policy, the office would have to give way on the strategy. This 1993 memo, with its emphasis on flexibility, foreshadowed the growing dissatisfaction with ECO, based partly on EPA's use of administrative discretion.

The issuance of EPA guidance technically moved the locus of implementation action from the national level to the state level. However, policy cycle stages and their substages do not follow neat sequential paths. The states and EPA were operating simultaneously on ECO policy. Even some companies in the affected areas were getting a jump on policy implementation by talking with state officials about plan development. Some companies already had the rudiments of ECO programs (e.g., carpool locator services, vanpools, and preferred carpool parking) and wanted to ensure that they received credit for them. They did not want these voluntary programs to raise the baseline from which they had to improve 25 percent.

Still, before measurable results could be achieved, states had to issue rules telling companies how to set up ECO programs. Some states wrote ECO programs into state statutes, whereas others promulgated regulations. To implement ECO according to the CAAA-90, states had to place legal, financial, and administrative obligations on businesses to change their employees' commuting behavior. States containing parts of one contiguous urban area did not all implement the same regulations. In some respects, the air quality regions that covered multiple states presented the greatest challenges for ECO implementation.

One Region: Four Regulations

The Philadelphia air quality region spreads into four states—Delaware, Maryland, New Jersey, and Pennsylvania. All four states initiated the lengthy process of developing ECO regulations soon after the law was signed and before EPA issued guidance. The individual states sent drafts of their ECO regulations to EPA regional offices for informal review before publishing official notices of their proposed actions in their state registers. All four states approached ECO differently and drafted different ECO rules. Since the ECO program was location-specific, employers in the Philadelphia region with facilities in more than one state were subject to multiple regulations.

Public Hearing: The Official Record

All four states in the Philadelphia region held public hearings during which oral and written comments were presented. They then considered the comments and sometimes published a summary of all the comments along with agency responses. A review of the formal record shows who was involved during this phase of implementation. It is probably not very surprising that the regulated community was overwhelmingly represented at these public meetings. Those who have something to lose are more vocal than those who have something to gain.

Maryland held three public hearings on ECO in July and August 1993. As Table 5.1 shows, the number of people attending and testifying was not very large. Businesses or their representatives (chambers of commerce) outnumbered any other group. Representatives of environmental groups were seldom present.

Table 5.1   Participation in ECO Public Hearings

Witness Type	Delaware	Maryland	New Jersey	Pennsylvania
Business	3	11	73	174
Government	0	2	21	19
Interest group	1	3	3	10
Citizen	0	2	5	3
Total	4[a]	18[a]	102[b]	203[b]

a. Oral comments only.
b. Includes written comments.

The content of the public testimony can be classified into two broad categories—substantive and procedural. Members of the business community expressed that their greatest substantive concerns were the costs of implementation and the loss of economic competitiveness by businesses in the air quality region. The substantive concerns of citizens who testified related to the goals of ECO.

More testimony focused on procedural and technical details of the proposed regulations than on substantive issues. This was largely due to the technical nature of the regulations but also due to the acceptance by many in the transportation subsystem that some form of ECO was inevitable. Businesses most frequently mentioned the burden of the required employee surveys. The next most frequently mentioned procedural issue related to the establishment of AVO zones.

Four public hearings on Pennsylvania's proposed ECO regulations resulted in the submission of ninety-four oral comments. A few representatives from environmental groups attended the Pennsylvania public hearing. They and other interest groups (those with a stake in the program's success, such as private consultants, transit operators, and alternative fuel and equipment suppliers) supported the program. Employers turned out in larger numbers, however, and their responses were strongly negative.

Public hearings in New Jersey and Delaware had approximately the same composition of witnesses. Private company representatives outnumbered other interests three to one. New Jersey had many witnesses, but only four testified in Delaware.

Comparing State Regulations

Although all four states in the Philadelphia air quality region worked from the same EPA guidance, they developed different regulations. Some differences were minor, but others were quite substantial. Table 5.2 summarizes the most important features of the four state regulations.

Legal Status

The legal status of the four state regulations provides some clues why the ECO programs in Pennsylvania and Maryland were suspended in 1994 and

Table 5.2   Employee Commute Options Regulations by State

	Delaware	Maryland	New Jersey	Pennsylvania
Target Populations				
Number of employers	400	1,700	1,000	2,500
Number of work sites	264	2,000	3,600	2,400
Number of employees	210,000	727,000	1,174,800	975,000
Legal Status				
Type	Statute	Regulations	Statute	Regulations
Regulations adopted	4/93	7/94	11/93	1/94
Regulations suspended				
prior to federal change	n.a.	5/95	n.a.	9/94
Implementing Agency	Transportation	Environment	Transportation	Environment
Program type	Plan review	Plan review with minimum measures	Plan review with performance standards	Performance standards
AVO Targets				
Number of zones	2	Floating	4	4
Target urban center	1.50	—	1.97	3.00
Target urban remainder	—	25% over base	1.73	1.75
Target suburban	1.45	—	1.46	1.58
Target rural/outer suburban	—	—	1.38	1.50
Fines, Penalties, and Enforcement				
Not registering	$1 per day per employee	None	$250–500/month	None
Not filing plan	$3 per day per employee	None	$1,000 per month	Agency penalties
Not implementing plan	$6 per day per employee	None	$5,000 per month	Agency penalties

1995, respectively, while those in Delaware and New Jersey continued through most of 1996. Pennsylvania and Maryland issued ECO regulations but did not put them in statutes; Delaware and New Jersey codified their regulations in state law. Federal policies codified in state laws are more likely to be implemented than those involving only state administrative decisions.

In Pennsylvania, ECO regulations were adopted January 28, 1994, by the Environmental Quality Board under the authority of the state's Air Pollution Control Act. Nine months later (September 1994), the state legislature voted to suspend ECO temporarily (until March 31, 1995), along with the unpopular emissions inspection program. Although Pennsylvania's Democratic governor vetoed the measure, the state legislature passed the suspension over his veto. The logic of the legislature in suspending ECO was based on the expectation that either EPA would reclassify the Philadelphia ozone nonattainment area from severe to serious or that Congress would amend CAAA-90 and make ECO voluntary. In February 1995, the state's newly elected Republican governor continued the suspension until the end of 1995. The legal status of the Pennsylvania ECO regulations was in limbo for more than a year until ECO was reformulated at the national level. Some Pennsylvania companies continued to implement programs (mostly the larger companies that had filed their plans before regulations were suspended), but most waited to see what would happen in the public arena.

Maryland published draft ECO regulations in July 1993 but withdrew the regulations two months later. After substantial input from the business community, the Maryland Department of the Environment adopted revised ECO regulations July 16, 1994. However, Maryland's ECO regulations would not become effective until October 15, 1995. Meanwhile, the Maryland Department of the Environment planned to begin a voluntary program in the hopes of working out any unexpected technical problems. Nevertheless, growing business opposition led the newly elected Democratic governor to suspend the ECO program in May 1995. The Maryland General Assembly followed suit by adopting an operating budget that eliminated all funds for ECO implementation.

The New Jersey state legislature codified ECO in a statute, "New Jersey Traffic Congestion and Air Pollution Control Act" (Public Law 1992, c. 32; N.J.S.A. 27:26-A-1 to 27:26-A-14). The law assigned responsibility for developing and administering the Employer Trip Reduction Program to the state's Department of Transportation. The law also established a deadline, November 15, 1996, by which employers had to achieve their target average passenger occupancy.

In November 1993 and again in May 1995, the New Jersey Department of Transportation adopted changes to the existing program: in both cases they went through the formal public comment period. The codification of

ECO gave the program an official and semipermanent status. It also meant that changing the program required action by the state legislature. The election of a new Republican governor in November 1993 did not substantially change the implementation of the program. In fact, even after Congress amended the CAAA-90 in December 1995 to make ECO voluntary, New Jersey's program continued to exist officially, and some companies continued to implement selected ECO program components. One reason ECO survived so long in New Jersey was its emphasis on mitigating congestion, a highly visible problem in the state.

Delaware's ECO regulations were adopted and codified in April 1993 (30 Delaware Code Section 2034 (3)). In July 1995 the law was amended to provide employers in Delaware's two affected counties (New Castle and Kent) the greater flexibility EPA had permitted. Moreover, the Delaware law included an automatic sunset provision that took effect if the federal law changed. Thus, Delaware's law automatically became voluntary when the federal law became voluntary. Delaware's ECO program was more flexible than New Jersey's and viewed by business as more friendly.

Implementing Agency

According to James Anderson, "A statute confers upon an agency only the legal authority to take action to implement policy on some topic" (1997, 229). The CAAA-90 gave the EPA and the states that legal authority. Authority aside, how effectively an agency carries out its legal mandate will be affected by the cooperation and opposition it encounters. As we have already discussed, EPA's political support at the federal level began to dissipate when a Republican Congress was inaugurated in 1995. For state agencies promoting ECO by pointing to EPA sanctions, the gorilla in the closet was turning into a monkey. The states' own support for ECO became more significant for the policy's survival, and this support directly related to which state agency was assigned the ECO program.

Maryland assigned ECO to its Department of the Environment (MDE), Air and Radiation Management Administration. This seemed like a perfect fit for a program designed to reduce ozone pollution. MDE, however, had no experience with transportation demand management programs and intended to supplement its staff with interagency reassignments from the Maryland Department of Transportation. These transfers never took place, and the program remained MDE's responsibility. The office was initially understaffed, with only two employees assigned to work on the development of ECO regulations. Five employees were added for ECO outreach efforts in 1994, but by then the Maryland program was on hold, and the state had a hard time convincing companies to make voluntary efforts.

In Pennsylvania, the Department of Environmental Resources (DER) developed regulations as the state's primary implementing agency. However, program management was shared with the City of Philadelphia, Department of Public Health, Air Management Services. The city took responsibility for Philadelphia County (which included the city), and DER assumed responsibility for the other four counties (Bucks, Chester, Delaware, and Montgomery), which made up the Pennsylvania portion of the Philadelphia air quality region. The DER assigned adequate staff to develop the regulations, and the Pennsylvania Department of Transportation cooperated by funding nine transportation management associations to provide training and assistance to employers developing ECO plans. The state received assistance from the Greater Philadelphia Chamber of Commerce to help organize an Employer Trip Reduction Program Coordinators' Council for outreach purposes and to give employee trip coordinators an opportunity to network.

Delaware and New Jersey assigned their programs to their state departments of transportation. Both states focused on the traffic congestion part of the law and assigned the program to a department with experience in transportation demand management. Both states already had extensive ride-matching programs in place before 1990. Although these programs were voluntary, they were state and federally funded.

In New Jersey, the Department of Transportation employed twelve people to administer the program, reflecting a strong commitment to congestion mitigation. Moreover, the state took a large credit for emission reductions (4.5 tons of nitrous oxides [Nox] per day) in its SIP, which also evidenced a belief that reducing vehicle miles would reduce air pollution.

Program Type

EPA's guidance outlined four acceptable options for state ECO programs. Delaware, New Jersey, and Maryland adopted the "plan-by-plan review" option. The Delaware program was simple, inexpensive, and employer-friendly. Delaware permitted employers to average, trade, and bank ECO credits. In addition, the state contracted with a transportation management association (TMA) to help employers comply with program deadlines.

The New Jersey program was the most complex and expensive of the four states. Employers had to meet strict guidelines for registration, develop employee surveys, and formulate and implement ECO plans. ECO plans had to be approved by an independent certifier before they were sent to the Department of Transportation for approval. Employers were required to meet deadlines and achieve the established targets. Steep fees for plan submission, stiff penalties for noncompliance, and fees for the independent certifiers increased the employers' program costs.

The review process in Maryland was much less complex than either

New Jersey or Delaware, but companies still had to register, submit plans, and file reports. Maryland had a set of minimum trip reduction measures (e.g., carpool locators), which every affected employer was required to implement. Maryland was the only state that used a "floating" AVO that allowed some employers to establish their own targets. However, employers were not required to meet their targets if they could show a good faith effort had been made.

Pennsylvania adopted the "performance standards" option that required employers to reach preestablished ECO targets. Pennsylvania's program was far less complicated than those of the other three states. Unlike the plan-by-plan review option in which the state reviewed and approved the company's plan before the company started its program, the performance standard gave the employer a free hand to do whatever it wanted as long as it achieved the target. The only requirement was conducting an employee survey to show compliance. The state would take action only when an employer failed to meet the target. Under this option, the knowledge, commitment, and efforts of the company were key to successful policy implementation. The state provided little assistance and made no assurances that company efforts would make a difference. If companies were committed to ECO's commute reduction goals, then effective implementation under Pennsylvania's program was possible, and the policy output of reduced solo commuting could be expected. However, commitment was not often forthcoming, and in fact, most companies were opposed to ECO and actively helped to get it suspended.

AVO Targets

The specifically mandated goal of increasing average vehicle occupancy (AVO) by 25 percent meant that some baseline had to be established from which to count the increase. Census data were generally used to establish the baseline and target AVOs for a region. EPA allowed a region to be treated as a single AVO zone or to be divided into several AVO zones. Thus, the number of AVO zones became an issue. A single AVO zone in a metropolitan area meant that the businesses in the city center might already exceed the target because many employees were already carpooling or riding the public transportation that effectively served the city center. Businesses in the suburbs, however, would have to make substantial change in commuting behavior and frequently were poorly served by public transportation. As the number of AVO zones increased, the target for city center businesses increased, and the target for suburban businesses decreased.

All four states adopted multiple zones, and all four used census data to determine the baseline AVOs for each zone.[4] Target AVOs were always 25 percent higher than baseline AVOs. States required companies to base their plans and reports on these numbers and required employee surveys to show how much change had to be made through the individual employer's program. Although the other three states had a small number of AVO zones and aggregated data across census tracts within each zone, Maryland adopted a unique floating AVO system that assigned employers the baseline AVO calculated for their 1990 census tract. If the company moved into the census tract after 1990, it could calculate its own baseline from its employee survey.

All four states made the baseline and target AVOs higher for the urban zones than for the suburban or rural zones. New Jersey had four zones, but the three highest AVOs applied only to that part of the state included in the New York City air quality region. New Jersey's lowest AVO applied to all the counties in the Philadelphia air quality region, whether urbanized Camden or rural South Jersey. Delaware had two zones with little difference between their target AVOs and encouraged businesses to move to the denser urban areas better served by public transit. Pennsylvania had four zones, with the city center AVO much higher than that of the outer suburbs. Maryland eventually had as many zones as census tracts, but had considered one, two, and three zones at various times during the process of developing regulations.

Fines, Penalties, and Enforcement

A public policy is only as good as its monitoring and enforcement. Otherwise it becomes what James Madison called a "parchment barrier" or what EPA staff called a "paper exercise." The growing complexity of ECO increased as lawyers for some affected businesses sought to clarify definitions, exemptions, and extensions. They wanted to find the easiest and most flexible program acceptable to EPA. By June 1995, ECO had become a paper exercise in some states. EPA had by then issued several memoranda adopting the "increased flexibility" recommendations of the Clean Air Act Advisory Committee. Our review of compliance provisions in the four-state ECO programs reveals important differences that could have affected the outcomes.

All four states recognized the need for a good faith effort and all four included such a provision. However, only Maryland and New Jersey specifically defined what a "good faith effort" was. All four states included some compliance criteria in their regulations, although occasionally they were minimally defined.

The states differed in exemptions, extensions, and appeals. New Jersey was the only state that allowed all three and required a $250 application fee for each request. Delaware also provided for an extensive appeal process for employers. Maryland, New Jersey, and Delaware lacked regulatory provisions for extensions, but Pennsylvania gave some companies extra time to file plans and conduct employee surveys.

As for fines and penalties, New Jersey and Delaware came down hard on the monetary side. Delaware did not list fees or penalties in its ECO regulations, but its fine schedule was clearly the most expensive, especially for a large employer. A large employer with 500 employees would be fined $500 per day for failing to register, $1,500 per day for failing to file an ECO plan, and $3,000 for not implementing a plan. Pennsylvania noted that administrative remedies would be used against those who did not file or implement a plan but did not specify what these remedies would be. Maryland did not address the issue at all. However, Maryland's plan was voluntary the first year and subject to later revision.

Our review of ECO regulations reveals significant differences among the four states in the Philadelphia region. These variations were probably big enough to produce different outcomes had the ECO programs been fully implemented. Yet it is possible that even the clearest and strongest regulations would not have produced the 25 percent increase in passenger occupancy expected within the four-year time line for implementation.

Conclusion

Policy implementation was once called the "missing link" in the policy process (Hargrove 1976). The complexities of turning an idea into an operable program are both challenging and frustrating. This stage has become even more complex in the 1990s as more federal policies require that state and local governments, the private sector, and individual citizens help implement them. Successful implementation requires the cooperation, coordination, and commitment of all the agencies in all levels of government involved. This requirement is often difficult to achieve.

ECO policy required four levels of implementation—the federal government to issue guidelines, the state governments to issue regulations, public and private employers to develop plans, and finally, employees to change their commuting behavior. Each of these four levels was critical to effective implementation of the program. In this chapter, we examined the first two levels, which necessarily affect the other two levels.

The economic development advocacy coalition emerged as an important force during ECO policy implementation, especially at the state level where many business interests are organized into associations (TMAs, chambers of commerce, etc.). The issuance of EPA guidance rather than regulations left the drafting of regulations up to the states and provided the economic development advocacy coalition with an opportunity to influence the development and content of these state regulations. Further, the delays by EPA in issuing its official guidance and further delays by the states in developing their final regulations gave business associations time to organize. The result was that most states changed their initial regulations to make them more flexible and acceptable to the affected businesses. The regulations promulgated by the four states in the Philadelphia air quality region initially contained wide variations that increased during ECO implementation.

Policy implementation has been described as an "up hill battle from start to finish," especially for policies that go against "strongly held beliefs, long-standing practices, and potent political forces" (Mazmanian and Sabatier 1981, 277). The implementation of ECO seems to fit this description. ECO lacked many preconditions for successful implementation:

- A policy must have clear and consistent objectives. Some states thought air quality was the primary objective and assigned implementation to their departments of the environment. Some states thought the primary objective was traffic congestion and assigned implementation to their departments of transportation.
- A policy must have an adequate causal theory. Not everyone believed in the link between commuting and air pollution. Only part of the traffic during the morning rush hour is due to commute trips by employees of large employers (Pisarski 1996). ECO did not affect non–work related driving.
- A policy must be designed to ensure compliance by implementing agencies. EPA had the authority to withhold federal highway funds from states who failed to promulgate acceptable ECO regulations. However, program implementation never reached this critical point. EPA was willing to compromise with states and promote flexibility, especially after the 1994 election.
- A policy must have committed and capable implementing officials. Federal officials disagreed with each other, particularly on how to handle pressures from states and affected employers. State agencies were frequently understaffed and faced opposition from state legislatures and governors.
- A policy must have the sustained support of interest groups. The quality of life advocacy coalition was instrumental in formulating ECO as part of the Clean Air Act but provided little support to agency staff to fend off assaults from the economic development advocacy coalition during the implementation stage.
- A policy will be affected by changes in the social and political environment. The election of a conservative, antiregulation Congress in 1994 made EPA and ECO easy targets to attack.

Now we turn from the public actors of implementation to the private actors, the businesses who had to implement ECO and the employees who were asked to change their commuting behavior.

Notes

1. To clarify, consider that the "simple" policy of revenue sharing (eliminated by Congress in 1986) involved the Office of Revenue Sharing, the Bureau of Accounts, the U.S. Postal Service, the Bureau of Economic Analysis, the Bureau of the Census, the U.S. Government Printing Office, and others (Edwards 1980, 3). In contrast, the interstate highway system, established in 1944 and scheduled for completion by the year 2000, is under the tight, vertical management of the DOT. The federal government pays 90 percent of the cost and supplies the standards; state highway departments pick up the rest of the cost and build the roads (following the strict standards established by DOT). When complete, the interstate highway grid will contain 44,000 miles of high-speed (at least four-lane) roads, approximately 1 percent of the roadways in the nation, but these roads will carry more than 20 percent of the traffic (Gerson 1997, 108–109).

2. States were given two years from the date the law was passed, November 15, 1990, to promulgate regulations requiring companies to develop ECO plans. Companies were given four years to develop plans, implement plans, and register results.

3. If a work site meets the target APO, vehicle mile credits equal zero. If a work site exceeds the target APO, vehicle mile credits is a positive number. If the work site fails to reach the target APO, vehicle mile credits is a negative number. A state may allow an employer to bank credits or trade them to another employer in the same nonattainment area. If a state does not allow banking or trading, it still may allow an employer with multiple work sites "to average compliance over all such worksites so long as individual compliance plans are submitted for each work-site" (USEPA 1992, 17).

4. Technically, AVO applies to an entire area and is calculated by the state for reporting to EPA. APO applies to an individual company and is the figure the company calculates and reports to the state. The term *AVO* is used in this book to avoid confusing the reader over what is conceptually the same thing.

6

Policy Implementation:
The Private Actors

Policy implementation is the most important and critical stage in the policy cycle. It is the "follow-through" of the policymaking process (Gerston 1997; Church and Nakamura 1993). Without implementation, policy remains an elusive goal. As we discussed in Chapter 5, implementation is a complex series of activities, due dates, and time frames. Most policy analysts adopt a top-down approach and focus on the "public" side of implementation—the decisions and actions of government bureaucrats. However there is a bottom-up perspective that analyzes the "private" side of implementation—the decisions and actions of the organizations and individuals who must change their behavior in order to implement, enforce, and apply the policy. Failure to take into account the private side of implementation and the linkages between the public and private sectors will result in poorly enforced and ineffective implementation, few policy outputs, little policy impact, and, ultimately, unsolved public problems. The key to policy success is a coordinated and committed effort by all levels of government to work with the private organizations and individuals who must make the changes required by the policy (Peters 1996). This chapter takes a bottom-up perspective and focuses on the private sector, the companies and employees affected by ECO. The companies were given the responsibility to implement and enforce ECO. The employees were asked to change their commuting behavior.

Willing or not, companies and their employees became important actors in the policy implementation stage since ECO directly targeted employers and indirectly targeted employees for behavior change. Although employers were specifically targeted by the ECO legislation, most were "latent actors" (Sabatier 1993, 24) during the earlier policy stages. The success of ECO depended upon how well employers with more than 100 employees could increase the average vehicle occupancy (AVO) among their employees. They faced major challenges in encouraging their employees to stop driving alone during rush hour.[1]

Implementing any policy places strains on the targeted organizations and individuals. They may be willing to comply when the required changes are small or incremental but unwilling to comply when major changes are required. Advocacy coalitions can greatly affect the willingness of targeted groups to comply with policy. Sometimes, targeted groups join an advocacy coalition to advance or stop a policy.

Advocacy coalitions play important roles in the implementation stage. Those successful at the policy formulation stage want to keep the policy moving forward. Those unsuccessful at the policy formulation stage hope to discourage compliance or change the policy and recover their lost ground. The involvement of opposing advocacy coalitions influences policy implementation. Sometimes a coalition that exerted little influence during an earlier stage will dominate the implementation stage. The role advocacy coalitions play during policy implementation varies depending on their size, wealth, organization, and cohesiveness. They can provide information, money, staff assistance, and symbolic support through letters and flyers.

Policy implementation is affected by more than just advocacy coalitions, however. Factors external to a policy subsystem influence policy implementation. (See Chapter 1.) Policies may be easier to implement during times of prosperity than during times of recession. They may be easier to implement following a crisis than when public attention is focused upon other matters. Further, target populations are not uniform, and some targeted individuals and groups find the policy easier to implement than others. Some individuals and groups have more resources than others and view the policy as consistent with their fundamental value system.

Compliance with Social Legislation

Legislation requiring businesses to respond to societal needs has proliferated since the late 1960s. This kind of legislation can be directed at the actions of the company itself or at the actions of company employees who are a captive audience of the general public. Policymakers frequently view companies as better able than government to change their employees' behavior. Companies already exert substantial control over employees through the hours and conditions of their employment. However, the ability of companies to control or influence employee behavior varies depending on the nature of the behavior. Behavior specifically related to employment (e.g., wearing safety glasses on the job) is easier for companies to influence than behavior outside the workplace (e.g., how the employee gets to work).

Companies are also more motivated to change employee behavior related to employment than employee behavior unrelated to employment.

Policymakers who make companies responsible for implementing policy must understand the factors that motivate companies to cooperate or not cooperate. We used the compliance with social legislation (CSL) model developed by Charles Greer and H. Kirk Downey (1982) to augment our bottom-up analysis of the "private" side of policy implementation. The CSL model is based on an earlier work by Kurt Lewin (1951), who hypothesized that the behavior of organizations is determined by a dynamic balance of countervailing forces. Greer and Downey call these opposing forces "driving forces" and "restraining forces." We prefer the terms "motivating factors" and "restraining factors," which we think more accurately describe their dynamic interplay. Motivating factors increase the probability of compliance behavior. Restraining factors decrease the likelihood of such behavior. (See Figure 6.1.)

Figure 6.1   Compliance with Social Legislation Model

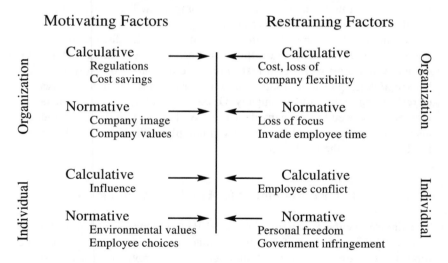

Source: Adapted from Greer and Downey 1982, 491.

The CSL model separates motivating and restraining factors into two dimensions. The first dimension identifies the company's decision criteria as either calculative or normative (Post 1978; Kangun and Moyer 1976). Calculative decisions are usually based on economic considerations and are pragmatic responses to specific situations (Greer and Downey 1982, 490).

A company assesses a situation and makes a pragmatic decision based on what will best help it reach its specified goals. Normative decisions are based on social and cultural values that are part of the company's belief system. A company makes normative decisions when it complies with a policy because it is the right thing to do or opposes a policy because it would infringe on the company's or its employees' rights.

The second dimension identifies the origin of the motivating and restraining factors for the organization or an individual. Company origins involve beliefs and values embodied in formal organizational policies, informal procedures, and corporate cultures developed over years of existence. Company origins are present when the organization responds in a predictable way no matter who the specific decisionmaker may be. At the same time, individuals who make company decisions may have their own interests and goals that affect their decisions. Individual origins display themselves when one person's decision differs from what another person within the same organization might decide. Individual decisions may even conflict with the organizations' values and goals. Analyzing both the organization's and individual decisionmakers' normative and calculative factors is important in understanding whether a company will comply with social legislation and act as an effective policy implementer.

Some factors can be either motivating or restraining factors, depending upon the circumstances. For example, cost can be a motivating factor if the company is doing well but a restraining factor if the company is facing bankruptcy. Employee attitudes and behavior can also be either motivating or restraining factors for a company. Of course, such attitudes and behavior may be influenced by company decisions. We believe that the CSL model is also useful in analyzing how employees comply with the social legislation implemented by their company.

State Encouragement for Employer Compliance

The Environmental Protection Agency (EPA) guidance said that each state "will establish a process of compliance plan submission, approval, periodic reporting on target achievement, and periodic plan revision" for companies (USEPA 1992, 4). Company plans had to "convincingly demonstrate" that the companies could achieve the necessary reduction in average vehicle occupancy (AVO). Companies faced varied regulations and assistance depending upon the state but also faced a definite schedule and many other common elements.

New Jersey had the most detailed and specific ECO regulations of the four states we studied. (See Box 6.1.) New Jersey companies had to use a standardized state-provided employee survey, submit the results on a stan-

---

Box 6.1 New Jersey Plan Requirements

1. Name, address, and work location of the company.
2. Title, signature, and telephone number of company's employee transportation coordinator.
3. Tabulated result of the latest and most recent annual employee survey.
4. A description of the physical and transportation service characteristics of the work location and demographic, work, and travel-related characteristics of the employee population.
5. A description of ECO strategies currently available to employees at the work location and additional ECO strategies the employer will implement at the work location.
6. Activities planned by the employer to implement the ECO program and a time schedule for implementation of the program
7. A description of the process by which the company will periodically monitor and review progress toward the APO target.
8. For update plans, a discussion of the company's "good faith efforts" to achieve the target APO, an explanation of why the ECO strategies included in the last plan did not produce the target, and a discussion of how the company planned to achieve the target within one year.

---

dardized form, and have plans certified by an independent certifier. Certifiers had to be trained by the state. Their job was to review company plans before they were submitted to the state in order to determine if the company could reach its target using the options selected. Companies could not use certifiers from the same organization that developed their plan. The other three states in our study did not require standardized surveys, standard plan submissions, or certifiers. Delaware, however, required some specific questions on the employee survey beyond what companies needed to develop their plans.

Sources of Information

The quality of life advocacy coalition seemed to take a back seat during ECO implementation. It did not actively promote ECO among companies or give companies information on how they could implement the policy. Federal and state agencies also made few efforts to sell ECO as a program that could eventually save companies money, boost employee morale, and improve their image as "green" employers. One EPA official lamented in retrospect that more time should have been spent educating people about ECO benefits before launching into the regulatory campaign (Ruth 1997).

In contrast, the economic development advocacy coalition was very active in the private side of the implementation stage. The U.S. Chamber of

Commerce and the National Parking Association, which had been marginally visible during formulation of the Clean Air Act Amendments of 1990 (CAAA-90), sent information to their members on the projected problems and costs of implementing ECO. They found natural allies among affected companies, transportation management associations, and local chambers of commerce. These groups became active participants in the coalition, increasing its resources and influence.

Most companies were unaware of ECO before the states began to implement the federal policy and were therefore unaware of the information on company implementation of transportation programs in California. During the formulation of CAAA-90, this information was orchestrated to project a positive picture of ECO-type programs and their impact on air quality and congestion. Further support had come from a few large companies elsewhere, such as 3M in Minnesota and Travelers Insurance in Connecticut, that had developed successful trip reduction programs motivated by company concerns.[2] By the time companies became aware of ECO, the economic development advocacy coalition had assembled other information showing that California was not achieving as much as previously proclaimed and that the self-initiated programs of a few large companies were not relevant to the situations of smaller companies.

Employee knowledge of ECO primarily came from the employer, and companies were uncertain how employees would react. Some companies expressed concern that implementing ECO would jeopardize relationships with their employees. If they pushed employees too hard, employees would refuse other company initiatives as well as ECO. In an era of employee empowerment and participatory management, company officials were concerned that telling employees to take public transportation and carpool would be viewed as an intrusion into their employees' private, nonwork life. Some employers were even concerned that they might be sued by employee groups. Therefore, rather than positively promoting ECO, most companies aligned themselves with their employees by blaming the state or the federal government for ECO, arguing that the "state is making us do this." This approach created an "us" (company and employees) versus "them" (state and federal government) mentality that allowed companies to implement ECO without alienating employees as long as the government enforced the policy.

State Assistance to Companies

State agencies varied considerably in the ways they helped companies implement ECO programs. Most states held a series of meetings and workshops during the initial development of regulations. Some states developed technical manuals and arranged for staff to visit companies personally and discuss company plans and needs with officials.

The amount and quality of state assistance given to companies depended upon which state agency was implementing ECO. Since the primary goal of ECO was cleaner air, many states assigned ECO to their departments of the environment. However, since the law specified a transportation solution to air pollution (i.e., trip reduction, mass transit use, carpools), other states assigned ECO to their departments of transportation. In our four-state study, Maryland and Pennsylvania assigned ECO to their environmental departments, whereas Delaware and New Jersey assigned it to their transportation departments (see Table 5.2). In general, transportation departments have more experience with and resources for helping companies deal with transportation issues than environmental departments. Thus, employers in states where ECO was implemented by transportation departments received more and better assistance from the government than those in states where environmental departments implemented ECO.

Time Frame

Most public policies require a long time to be fully implemented. Public behavior often changes slowly even when external system factors are favorable and changes even more slowly when external system factors are not favorable. Despite this, public policy often specifies short implementation schedules. People seem to have an insatiable need to get immediate answers to the ever-present policy question, "Did it work?" or "Has the problem been solved?" Elected officials need tangible evidence of policy success before their next election (Anderson 1997). Quick implementation may prevent opposing advocacy coalitions from mustering the resources necessary to attack a policy. A fast implementation schedule can be both a restraining force and a motivating force for company compliance with social legislation. Companies are motivated to move quickly and not question the validity or logic of the policy. However, they are restrained by lack of information and the time necessary to plan a comprehensive strategy.

The CAAA-90 established a tight schedule for ECO implementation. Several observers noted that expecting companies to change the driving behavior of their employees in just a few years was unrealistic (Adler 1994).[3] The first deadline for companies was November 15, 1994, when they had to file their initial ECO compliance plans. Companies then had just two years to implement their plans and achieve the AVO target goals averaging a 25 percent reduction in solo commuting trips by their employees.

The formulated schedule required a tight sequence of events. All actors in the process had to do their respective jobs in a timely fashion. The first delay occurred when EPA's final guidance was issued one month after state implementation plans (SIPs) were due. Although EPA gave the states no "official" extensions in submitting their SIPs, EPA could hardly expect states to complete their regulations without the final guidance. Eight of the ten affected states failed to meet the November 1992 deadline for submitting their ECO compliance plans to EPA.[4]

This interruption of the momentum and energy that usually accompanies the passage of a new law gave the economic development advocacy coalition time to organize affected companies against ECO. Companies started putting state agencies on the defensive. Agencies initially saw ECO as a way to address some difficult problems they faced and promoted the quality of life advocacy coalition's pro-ECO position. As anti-ECO pressures mounted, state agencies moved to the role of policy broker and passed the blame onto the EPA for failing to give them timely guidance. New Jersey, the most compliant of the states we studied, expressed its resentment of EPA in a response to a question from a concerned company:

> The ETR [employer trip reduction] program was developed in conjunction with the New Jersey business community but under the constraints of very limiting prescriptive USEPA guidance. As a result, the program placed unprecedented pressure and cost upon businesses, employees and their families. USEPA has offered relief to the states in the form of additional flexibilities. (NJR 2437 1995)

Such state responses showed a weakening of the influence of the quality of life advocacy coalition compared with that of the economic development coalition. Some companies decided they did not have to do anything.

Transportation Options

ECO gave companies no new transportation options to help them implement ECO. Most transportation components—streets, parking, transit, and zoning—are under the control of local governments. Local governments are also the logical entities to administer areawide carpool programs, establish park-and-ride lots, develop bicycle paths, and put in sidewalks. Local governments are very interested in both the quality of life for their citizens and the economic development that funds that quality of life, but they were not included in ECO. ECO neither required local government involvement nor provided incentives for local governments to expand transportation options. Therefore, companies could count on little transportation assistance from local or regional public agencies. Explicitly including local governments in ECO would have created another implementation "check point" (Mazmanian and Sabatier 1981, 1983), and yet it would have given these key transportation and urban actors a stake in ECO's success. ECO, however, treated local governments only as employers, subject to the same regulations as the private sector. This allowed them to be recruited by the economic development advocacy coalition along with private employers. Thus, the lack of any new transportation options from local governments was a restraining factor for companies trying to comply with ECO, rather than an internal motivating source of additional information and new tools.

Legal Tools

ECO gave companies no new legal tools as motivating factors for change. In fact, some existing laws restrained implementation of ECO. One tax law allows employers to offer free parking to employees as nontaxable job benefits and write off the whole cost of the parking against the companies' taxes. Monetary transit subsidies, however, are treated by the law as taxable employee income; companies can claim them as regular business expenses only up to a certain amount. Thus this law encourages companies to give their employees free parking but not bus passes. Since it has been well documented that free parking or parking rates below the market price contribute to solo commuting (Shoup and Pickrell 1980; Surber, Shoup, and Wachs 1984; Willson and Shoup 1990; Brownstone and Golob 1992), this law functions as a restraining factor on company implementation of ECO.

The Fair Labor Standards Act creates another restraint. The act, passed in 1974 to ensure that employees were treated fairly, requires that hourly wage employees be paid overtime when they work more than forty hours in a seven-day period. The law does not permit the averaging of hours over two or more weeks. Companies exploring a compressed work week of nine, mostly nine-hour, days over a two-week period discovered they would have to pay overtime for the extra hours worked during the first five-day week.[5] This financial restraining factor frequently offsets the motivating factor of employees' strong preference for regularly scheduled three-day weekends. Although compressed work weeks offered large clean air benefits, even if employees continued to commute alone (assuming employees did not use their cars to run errands on days off), this stable external system factor restrained rather than motivated use of compressed work weeks.

Costs

Companies did not know the direct and indirect costs of ECO programs. This uncertainty became a major restraining factor. Studies reported savings as high as $533 per employee per year and costs as great as $750 per employee per year (COMSIS 1994; Ernst and Young 1992; Giuliano and

Wachs 1992; Green 1995; Stewart 1994). The average cost, however, was around $100 per employee per year. The applicability of these cost estimates to other companies is difficult to assess since many of these studies were based on nonrandom samples of companies. Even the random, systematic studies done in California found it difficult to separate a company's capital expenditures on facilities tangentially related to ECO, such as a new cafeteria, gym, or daycare facility, from direct ECO costs. Further, most cost data were based on self-reporting by companies that had a tendency to overreport expenditures and underreport benefits. One researcher found that nearly 70 percent of the money reportedly spent to promote ride sharing was used for program administration rather than for program incentives (Green 1994).

An opposite motivating factor, although largely unknown, was the cost of noncompliance. Companies expressed concern that they could spend a lot of money implementing ECO programs and still be slapped with fines because the programs did not achieve the desired outcomes. They clamored for clauses within regulations that gave credit for "good faith efforts." EPA soon made it clear that it would not penalize any state that developed regulations and made good faith efforts to get companies to implement ECO, even if they did not achieve their target AVOs. All four of the states in the Philadelphia air quality region, in turn, gave credit for good faith efforts of companies, although only two defined the term in their regulations. Only New Jersey developed a detailed schedule of financial penalties for companies that failed to make these good faith efforts.

Company Compliance

Sample of Companies

Our study included a sample of companies in the Philadelphia air quality region that would be affected by ECO. We randomly sampled from within thirty-six strata, formed by classifying companies into four geographical groups (City of Philadelphia, Pennsylvania counties, Delaware counties, and New Jersey counties), three size groups (100–499, 500–999, and 1,000 or more employees), and three industry groups (manufacturing, service, and other). We supplemented this basic sample with two additional samples. The first supplement included a set of Delaware companies not included in the basic sample. A second supplement included Maryland companies, primarily in the Baltimore air quality region, that had indicated that they intended to implement voluntary ECO programs in anticipation of the delayed state regulations.

We contacted 497 companies in an attempt to recruit three companies in each of the thirty-six strata of the basic sample. About half of those we contacted did not have a site in the region with 100 or more employees, no longer had a telephone listing, or were otherwise not appropriate for our study. A total of 112 companies agreed to participate in the research, 41 percent of the companies we contacted that met the criteria for our study. (See Study Participation: Companies and Table A.1 in the Appendix for more details.)

Company Interest in ECO

Recruitment of companies for the study began in June 1994 and continued until July 1995. During the early months of recruitment, most of the companies agreed to participate in our study. (See Figure 6.2.) After Pennsylvania suspended its program in November 1994, about one-third of the contacted companies agreed to participate.

Figure 6.2   Percentage of Companies Agreeing to Participate in the Evaluation by Month of Contact

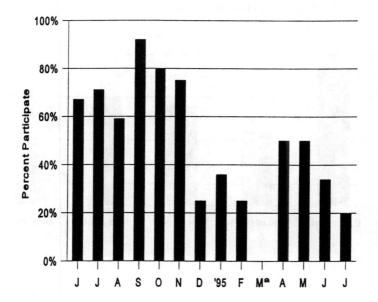

Note: a. Only one company contacted.

The CSL model helps us understand company willingness to participate in our study as well as compliance with ECO legislation. We offered to conduct and analyze ECO employee surveys without charge as a motivating calculative factor for participation. We also offered to provide information about what other companies in the study were doing and the most cost-effective ways other companies had found to implement ECO. For a motivating normative factor, we stressed the value of participating in university research and advancing knowledge. However, participation in our study also had some restraining factors. Our survey form was longer than the ones required by the states, and additional production time could be lost as employees completed it. Some companies thought our longer survey invaded employees' privacy too much, a restraining normative factor.

State implementation activities affected whether motivating factors outweighed restraining factors. The highest study participation came from suburban Pennsylvania companies recruited early in our study, and the lowest participation came from Delaware companies recruited later in the study. (See Figure 6.3.)

Figure 6.3   Percentage of Contacted Companies Participating in the Evaluation, by State

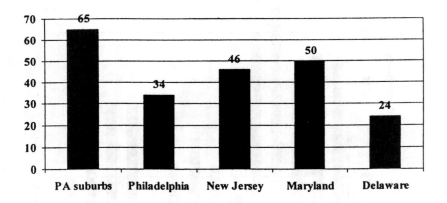

Pennsylvania required companies with 1,000 employees or more to submit ECO plans by November 1994, with the rest submitting plans by November 1995. The Pennsylvania Department of Environmental Resources therefore focused most of its early efforts on providing information to large companies. When we began contacting Pennsylvania companies in the summer and fall of 1994, many smaller Pennsylvania companies did not know about ECO before our contact or had little information about it. At that time, Pennsylvania's ECO regulations were very demanding, requiring that the target outcomes be met with little guidance on how companies could achieve them or even what would be an acceptable plan. Few small companies had personnel they could spare to learn about ECO, and few had internal resources for conducting and analyzing surveys. These small Pennsylvania companies, concentrated in the suburban counties, saw substantial benefits from participating in our study. Large Pennsylvania companies concentrated in Philadelphia, however, saw less benefit from participation in our study. They had already received ECO information from the state. Most had the internal capacity to conduct employee surveys, and many had already done so. We made few attempts to recruit Pennsylvania companies after the Pennsylvania legislature suspended the state's ECO program, and we stopped altogether in January 1995 after the governor made the suspension permanent.

Delaware companies had the least motivation to participate in our study, and only one-fourth agreed to do so. Delaware started registering affected companies in April 1994 and immediately started providing information and survey assistance. The director of Delaware's transportation management association (TMA) was a member of the state legislative committee that had developed Delaware's regulations. The TMA kept most of Delaware's companies up to date on ECO and helped many of them conduct employee surveys before our initial contact with them. Additional factors reduced the motivation for companies to participate in our study: the state-approved employee survey form was much shorter than our form and did not collect employee identification; the state required several questions not on our survey, which meant employees would have to complete both survey forms.

New Jersey and Maryland companies had intermediate motivating factors to participate in our study. New Jersey was just starting to require companies to register and develop plans at the time we began recruiting for our study. It also required employers to have their survey results and plans certified by accredited professionals who had not been involved in conducting the survey or in preparing the plans. However, other factors restrained company motivation. New Jersey gave companies substantial information, survey forms, and plan formats on computer diskettes. It also required companies to use the state's employee survey form, with supplemental questions allowed only at the end. Our survey form had to be treated as a supplement rather than a replacement and, therefore, was considered too time-consuming by most companies.

Maryland had not completed its regulations by the time we started recruiting, and so companies had no incentive to participate in our study. However, Maryland had promised to look favorably on companies that voluntarily started ECO programs, and companies starting early had more time to achieve their targets. Maryland also accepted our employee survey as a model for the eventual state survey.

Although organizational calculative motivating factors to comply with social legislation can be seen in responses to our study recruitment, individual calculative motivating factors also seemed to be involved in the recruitment process. About two-thirds of the companies assigned responsibility for ECO to their directors of human resources, who are responsible for matters relating to employees. Human resource directors saw value in telling employees that the survey was being done by outsiders (a university) rather than the company. Thus, the company could not be accused of prying into the private lives of employees or of considering the programs mentioned in the survey. These directors also knew it would take them time to learn about ECO and to conduct and analyze employee surveys. One human resource director expressed another individual motivating factor for his company's participation. He had never been allowed to survey employees, and he could learn a lot about them from our survey. If Pennsylvania had not suspended its ECO program, his individual calculative motivating factors might have overcome the organization's normative restraining factors, namely, outsiders gaining information about the company. Other human resource directors saw ECO as an opportunity to push for schedule flexibility (flextime, compressed work week, telecommuting), which would make employees happier and reduce the number of problems seen by the human resources department. These individual motivating factors, however, were generally not sufficient to overcome organizational restraining factors after the state requirements were gone.

Implementing ECO Plans

Our study obtained information about ECO compliance from ninety-one of the companies that initially agreed to participate. About one-third of these companies never developed an ECO plan, about one-third developed a plan and implemented some part of it before abandoning it, and about one-third developed a plan and expected to continue implementing their ECO plans during the year following the survey, regardless of state regulations. However, these proportions differed greatly by state. Most of the companies in Delaware and New Jersey developed a plan, compared to less than one-third of the companies in Maryland and Pennsylvania. (See Figure 6.4.) Some companies planned ECO programs to meet state requirements

Figure 6.4   Percentage of Companies with ECO Programs

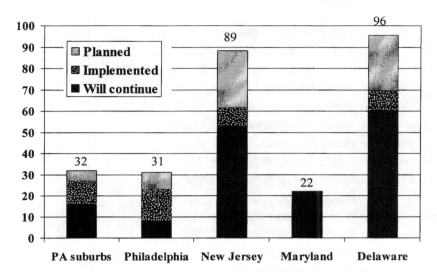

but never implemented any part of the plan. Others implemented ECO programs (or were in the process of implementing at the time of the survey) but would stop them if not required to continue. The majority (52 percent) of companies in Delaware and a large percentage (39 percent) in New Jersey, however, said they would continue implementing their ECO plans regardless of state regulations. Very few Pennsylvania companies (8 percent in the city and 16 percent in the suburbs) planned to implement ECO programs in the absence of regulations. All Maryland companies in our sample had indicated they would initiate voluntary ECO programs in advance of any state regulations. One-fourth followed through; the rest never completed a plan once they realized ECO would never be required.

Our company survey included questions about twenty-one possible components of ECO plans. (See Figure 6.5.) Companies in Delaware and New Jersey planned to implement about one-third of the possible components (average of 6.8 and 6.0 respectively), with some companies planning as many as fifteen components. Two-thirds of the companies planned a ride-matching component to meet ECO requirements; half planned to guarantee a ride home for employees who missed their carpool or transit

Figure 6.5   Percentage of Companies Including ECO Components in Plan

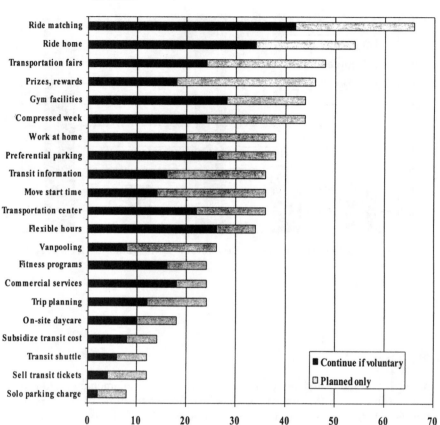

connections; and more than 40 percent planned transportation fairs, prizes, gym facilities, and compressed work weeks. Very few companies planned to use more costly measures such as daycare, transit subsidies, or transit shuttles. Charging for solo parking was the least likely component, despite the evidence cited earlier that this measure would have the biggest impact on solo commuting.

State regulations provided both calculative motivating and restraining factors for companies to implement ECO programs. One-fourth of the companies that implemented ECO plans explicitly said they did so because of state regulations. (See Box 6.2.) Most of those who did not implement any ECO program said they were not sure they had to or believed the state

---

**Box 6.2   Reasons Given by Companies for Implementing
and Not Implementing ECO Programs**

Implementing
- State requirement (24 percent)
- Employees benefit or want it (22 percent)
- Financial benefit to company (14 percent)
- Company image (12 percent)
- Right thing to do (10 percent)

Not Implementing
- No or uncertain state requirement (32 percent)
- Too expensive (10 percent)
- Too much trouble (6 percent)
- Already exceeded requirements (6 percent)
- Can't change employee behavior (4 percent)
- Too intrusive on employees (2 percent)

---

would not enforce ECO. Taken together, more than half the companies surveyed cited either the presence or absence of state regulations as a primary factor in their decision to implement or ignore ECO.

Beyond state regulations, companies gave very different reasons for implementing or not implementing ECO programs. Implementing companies cited employee benefits (22 percent), financial benefits (14 percent), and company image (12 percent) as reasons. Few of those ignoring ECO cited employee-related reasons. Some suggested, however, that employee attitudes and expectations could be restraining factors. ECO surveys and planning might raise employee expectations or fears, with subsequent negative consequences for companies. For example, one company found that employees wanted on-site daycare, and so the company engaged a provider. When it could not convince the industrial park management to build or lease suitable daycare space within the industrial park, employees became unhappy with the company. Several companies reported concern that employees would view some ECO programs (e.g., schedule changes, transit subsidies, and preferential parking for carpoolers) as new employee benefits that would be hard to stop later if the law changed.

Companies said they were likely to continue the two most frequently planned components, ride matching and guaranteed ride home, whether or not the states required them to implement or continue ECO programs. (Refer to Figure 6.5.) Schedule changes were the next most likely ECO components to be continued. One-fourth of the companies planned to continue flexible work hours, even without state requirements. One-fourth also planned to continue compressed work weeks, which was more than half of those that included compressed work weeks in their ECO plans.

Company Characteristics

We developed two summary measures of compliance with ECO legislation, planning compliance and overall compliance, to test the CSL model. Planning compliance is measured by the number of components that companies included in their ECO plan. Overall compliance is measured by the extent to which ECO programs were implemented by companies. (See Analytic Measures: Companies in the Appendix for details.)

Multiple regression analysis showed that half of the variation in planning compliance and overall compliance can be explained by three factors: state location, company size, and TMA membership. (See Appendix Table A.2.) Most important is location. New Jersey companies included six more components, and Delaware companies included four more components, in their ECO plans than did companies in Maryland and Pennsylvania. New Jersey and Delaware companies also scored significantly higher on Overall Compliance than did Maryland and Pennsylvania companies. This finding suggests that the more detailed the state regulations, the stricter the codification of regulations into law, and the longer the regulations are in place, the more likely companies are to comply with policies like ECO. Once companies develop programs because the state requires them to, they are likely to continue the programs even if the motivating factor of state law is removed.

Company size was the second factor for ECO compliance. Companies with 750 or more employees planned two more ECO components on average than did companies with less than 750 employees. Larger companies were also more likely than smaller companies to continue ECO programs without state regulations. Planning and administering programs cost time and money whatever the number of employees. Larger companies can spread these costs over more employees, thus reducing the relative costs of planning and implementing social legislation. Since the restraining factor of cost is lower for larger companies, they are more likely to comply with social legislation than are smaller companies.

Membership in a TMA was the third significant factor associated with ECO compliance. Companies belonging to TMAs planned 2.5 more ECO components than did companies not belonging to TMAs. They also were more likely than non-TMA companies to continue their ECO programs without state regulations. TMA membership is both a calculative and a normative motivating factor. TMAs provide information and services to member companies that reduce the costs of learning about regulations, of planning, and of implementing transportation-related programs. TMAs have traditionally been sources of transit and carpool information and have experience in administering carpool programs. Membership in TMAs also reflects a company's belief that it is appropriate to get involved in commuting issues. Finally, a company's membership in a TMA could mean that it received its initial information on ECO from the TMA at the time the TMA viewed ECO positively as a way to increase TMA membership and services. (Some TMAs later coordinated opposition to ECO when they found their members opposed to ECO.)

We tested a number of additional factors that might be related to ECO compliance. One was the position of the transportation coordinator in the company since individual calculative and normative factors might be greater for some company officials than for others. Our initial recruitment experience suggested that human resources executives would be more open to ECO programs than other company officials. However, the position of the transportation coordinator within the company's organization made no difference in whether the company developed a plan or in its overall compliance. Moreover, the survey data showed that human resource directors planned two fewer components than did other types of company officials, when the company did plan an ECO program. This suggests that if individual factors did come into play, they were a function of the characteristics of the individual rather than the position of the decisionmaker within the company.

Although Box 6.2 shows that some companies planned to continue ECO programs for image reasons, the importance of their overall image or their "green" image had no relationship to company compliance with ECO. The type of entity—publicly traded for-profit company, privately held for-profit company, or nonprofit organization—also did not matter. Compliance was not affected by the site's being the company's headquarters or a branch location. Manufacturing, service, and other types of companies responded to ECO in much the same way. Union and nonunion companies responded similarly, as did companies operating different numbers of shifts each day and different numbers of days during the week. Even the proximity of public transportation did not appear to affect company response to ECO.

ECO Program Costs

A major calculative factor is cost. This factor was cited in our study more frequently as a reason for implementing ECO than for not implementing ECO. We obtained personnel and direct planning costs for twenty-three companies. Personnel costs included attending meetings on the regulations, communicating with management and employees, surveying employees, writing ECO plans, and marketing the plans to employees. Direct costs included surveying employees for baseline data, hiring consultants to prepare and certify plans, installing bike racks, and modifying parking areas for carpools. Companies spent an average of $24,000 during their planning process, although half of them spent less than $14,000. (See Table 6.1.) This averaged $71 per employee, with half the companies spending less than $31 per employee. We expected higher planning costs in New Jersey than in the other three states due to its strict requirements, which included company-paid certification before submitting plans. New Jersey companies did report substantially higher median planning costs than did companies in the other three states, but lower mean planning costs than Delaware companies and lower planning costs per employee than Delaware and Maryland companies. The mean planning costs per employee for reporting Maryland companies were at least twice as high as in other states, and the median costs per employee were higher than in two of the other three states, suggesting that planning to implement voluntary policy is at least as expensive as planning to implement required policy.

Table 6.1   Total and Per Employee Planning Costs by State

State	Number of Companies	Total Spent		Cost per Employee	
		Mean	Median	Mean	Median
NJ	7	$28,000	$22,000	$39	$28
DE	8	$36,000	$11,000	$71	$44
PA	4	$23,000	$18,000	$18	$16
MD	4	$22,000	$5,000	$160	$34
Total	23	$24,000	$14,000	$71	$31

The large companies in our study spent more for planning than small companies. However, there appears to be a minimum cost for all companies of about $20,000 for planning an ECO program, with costs increasing after about 500 employees. (See Figure 6.6.) Relative planning costs, however, decreased with size from $145 per employee for companies with 100–199 employees to only $22 per employee for companies with 500 or more employees. Large companies invested more total dollars in planning ECO programs but spent less per employee than did small companies.

Fewer companies implemented ECO programs than planned them. Both planning and implementation costs were available for only six New Jersey companies. These six companies spent an average of $37,000 to plan their program and expected to spend an average of $42,000 per year to implement their programs. The companies varied greatly by size (from 103 to 4,406 employees) and in the amount they spent per employee for

Figure 6.6 Average Cost per Employee and per Company, by Size of Company

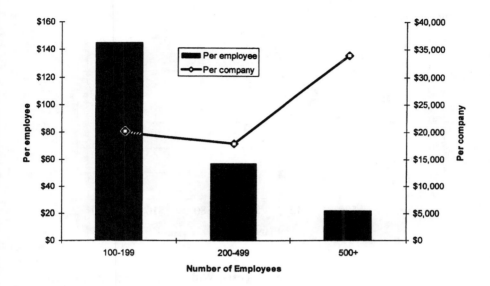

planning their ECO programs (from $9 to $109 per employee). (See Figure 6.7.) They varied much less in the annual amount they expected to spend per employee to implement their programs, a range of $10 to $40, with a median cost of $30. Only two companies (those above the diagonal line) expected their yearly implementation costs to exceed their initial planning costs. The other four companies reported planning costs that were two to five times greater per employee than their expected annual costs of implementing ECO. None of our companies anticipated any savings from ECO, but their average expected costs were at the lower end of the range of costs found in other studies. These data suggest that the cost of implementing ECO would not have been a major restraining factor for company compliance, if companies knew at the beginning what they would be. However, the economic development advocacy coalition used the unknown amount of the costs as its major argument against ECO. The quality of life advocacy coalition missed an opportunity to promote company implementation of ECO by not providing information that would reduce planning costs and

**Figure 6.7   Past Planning Costs and Expected Annual
Implementing Costs per Employee for Six
New Jersey Companies**

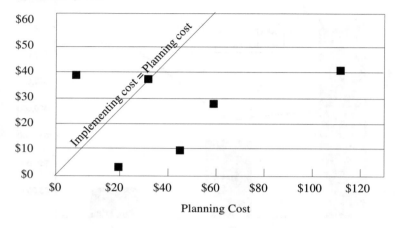

Planning Cost

the uncertainty about implementing costs. Without state regulations, companies had no motivation to invest the amounts necessary even to plan compliance. Only those that had incurred the planning costs before the states made ECO voluntary continued to implement ECO.

Likelihood of Change in Commute Behavior

Companies can offer many different types of programs to change employee behavior, but the success of the programs is likely to be affected by employee attitudes toward them. Our survey of 5,000 employees in fifteen companies asked about the likelihood that they would be influenced by commute reduction programs that their companies might adopt. We also asked employees how they felt about various negative actions companies might take to meet their ECO obligations, although EPA specifically forbade companies from imposing sanctions against employees who continued to commute alone. What employees say would influence them is not necessarily the same as what they would actually do, but attitudes do give insight into the types of ECO programs most likely to succeed. Two-thirds of the employees responding to our study said that a program involving a compressed work week would be very or somewhat influential. (See Figure 6.8.) More than half of employees said telecommuting, guaranteed ride home, and flexible work hour programs would be very or fairly influential.

Figure 6.8   Percentage of Employees by Potential Program Influence

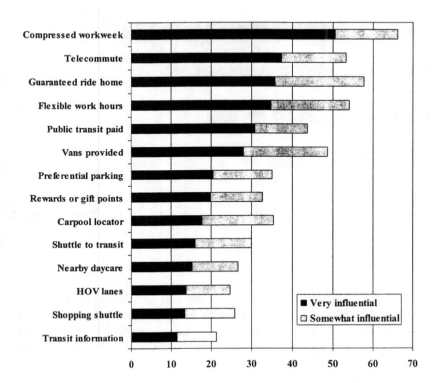

One-fifth said programs that provided transit information would be influential. The fourteen possible programs clustered into three factors, from which we created three scales: schedule change programs, carpool programs, and transit programs.[6] (See Analytic Measures: Employees and Figure A.2 in the Appendix for more details.)

Negative sanctions by companies were much less acceptable to employees than positive promotions. Few employees thought any sanctions were justified, even in the face of a fine from the state or loss of federal contracts. They were most likely to feel that moving the company closer to public transportation would be justified, but this was acceptable to only one in six (16 percent) employees. Only one in twenty (5 percent) felt companies should be able to make employees move closer to work. The six questions involving negative sanctions clustered into a single summary scale.[7]

Schedule change programs had the most potential to influence employee commute behaviors, with an average influence score of 3.5 on a range of

Figure 6.9   Potential Program Influence

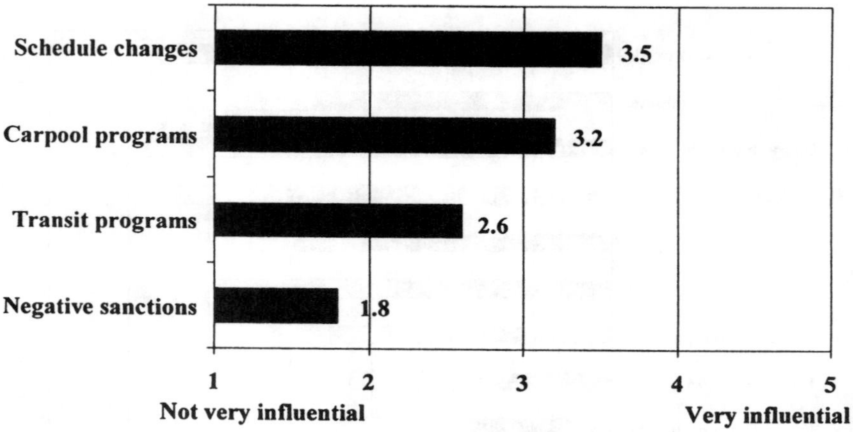

1 to 5. (See Figure 6.9.) Carpool programs follow next, with an average influence score of 3.2. Transit programs were even less likely to influence employees, with a 2.6 average score. The average negative sanctions score of 1.8 indicates this approach to implementing ECO is least supported by employees, even though the scale values have a slightly different meaning than those for the positive programs.

Schedule Change Programs

Employees say they would be motivated most by schedule change programs: flexible scheduling of work hours, compressed work weeks with longer hours on fewer days, and telecommuting. As discussed earlier, compressed work weeks and flexible hours were among the four ECO program components most likely to continue even after ECO was made voluntary. Schedule change programs would affect all employees about equally. Only 15 percent of the variation in how employees responded to the schedule programs can be explained by their attitudes, background characteristics, and behaviors. (See Table 6.2.) Schedule changes would most influence employees already positive about carpooling, with high ideals for teamwork, and sensitive to the environment. Women, older employees, and higher income employees are slightly more likely to be influenced by schedule change programs than are men, younger employees, and those with lower family incomes. Access to cars, commute distance, education,

**Table 6.2   Regression of Likely Influence of Company Programs
to Reduce Rush Hour Commuting on Current
Behaviors, Attitudes, and Background Characteristics**
(standardized regression coefficients)

Independent Variable	Schedule Programs	Carpool Programs	Transit Programs	Negative Sanctions
Solo commuting	0.03	−0.04	−0.10**	−0.04
Environmental actions	0.04	0.05*	0.04	0.05*
Outside relationships	0.03	0.02	0.08**	0.08**
Current teamwork	−0.07*	−0.05	−0.03	0.05
Office relationships	0.03	0.06*	0.01	0.02
Carpooling orientation	0.21**	0.34**	0.28**	0.11**
Environment sensitivity	0.08**	0.17**	0.10**	0.02
Man over nature	−0.04	−0.01	0.03	0.08**
Ideal teamwork	0.15**	0.12**	0.09**	0.01
Cars per household member	0.01	−0.03	−0.06**	−0.03
Education	0.03	0.04*	−0.00	0.05*
Age	−0.14**	0.00	−0.08**	−0.00
Family income	0.09**	0.00	−0.10**	−0.02
Commute distance	0.02	0.03	0.02	0.01
Male	−0.08**	−0.05**	−0.04*	0.02
R^2	0.15	0.25	0.21	0.05

* $p < 0.01$ ** $p < 0.001$

current commuting, and current environmental behaviors are unrelated to
the influence of schedule programs.

Carpool Programs

Carpool programs, frequently planned by companies and somewhat likely
to continue under a voluntary ECO, were the second most likely factor to
motivate employees to change their commuting behavior. Predicting which
employees would say they would be motivated by carpool programs is
easy. The more positively employees already view carpooling, the more
sensitive they are to the environment, and the more highly they value teamwork, the more likely they will be positively influenced by company-
promoted carpool programs ($R^2 = 0.25$). Neither employees' current commuting behavior nor the number of cars per household member affects the
influence of company carpool programs, once the three attitudes have been
taken into account. Women and those with more education are slightly
more likely to say they would be influenced by carpool programs than are
men and those with less education. Employees who currently engage in
environment-related behaviors (e.g., recycling and supporting environmen-

tal groups) and interact more with coworkers at the work site are more likely to be influenced by carpool programs than employees who are less active environmentally or socially.

Transit Programs

Transit programs would motivate employees less than the other two positive programs, and companies were least likely to plan or continue them. It is generally current transit riders and those with few cars per household member who say they would be influenced by transit programs. Again, attitudes are the best predictors of influence. Employees who view carpooling more positively, who are more sensitive to the environment, and who value teamwork more highly are more likely to be motivated by transit programs to change their commuting patterns. Employee background characteristics, though, do affect the likely influence of transit programs. Women, younger employees, those with lower family incomes, and those with fewer cars per household member are more likely to be influenced by transit programs than are men, older employees, those with higher family incomes, and those with more cars. Current commuting behavior does not have any additional influence, but greater environmental activity depresses the likely influence of transit programs.

Negative Sanctions

Negative sanctions were least liked by employees, their influence was least predictable, and almost no companies planned to implement any. Only 5 percent of employee variation in the influence of negative sanctions can be explained by the factors considered in this study. (See Table 6.2.) Again, however, attitudes explain the most. The more employees are oriented toward carpooling, the more likely they are to think companies would be justified in taking actions to restrict solo commuting. Employees who believe humans can and should control nature are more likely to feel their companies would be justified in forcing behavioral changes through negative sanctions. More frequent social interactions with other employees outside the workplace and more frequent environmental actions also increase the feelings that the company would be justified in imposing negative sanctions.

Company Experiences

Projections about the likely influence of any program may be different from the actual change resulting from that program. In the case of ECO, anecdotal information on changes in employees commuting behavior was available from some companies. Some employees switched to carpools, but most employees' behavioral change involved when they arrived at work. Employees liked changing their work schedules better than joining carpools or riding public transit. As one employee clearly expressed, "You can't take my car away from me!" Although information is sketchy, our research yielded several accounts of how different companies implemented ECO.

Company A, a manufacturer on the edge of a small town in southern New Jersey and unreachable by public transportation, decided to meet ECO requirements by changing the start of the morning shift from 6:00 A.M. to 5:30 A.M. One-fourth of its labor force made the change, which increased its AVO sufficiently to exceed the target set by the state. However, this change had no impact on how employees commuted to work. Employee surveys showed that only 8 percent of the employees commuted differently after the change than they did a year earlier before the company implemented its ECO program. Half of these (4 percent) changed from driving alone to carpooling or from carpooling to walking or riding a bicycle. The other half did the reverse and changed from carpooling or walking to driving alone. When the state decided not to enforce ECO, the company returned the normal shift start time to 6:00 A.M. Half of those who moved to the earlier start time returned to driving during the morning rush hours defined by ECO.

Company B, a service company in suburban New Jersey, had a large professional workforce of 2,400 employees. Some of its operations ran twenty-four hours a day and seven days a week. Company B started working on ECO in December 1993, organized a planning team in February 1994, and started an educational process. When the company's planning team started, they felt ECO programs would never work in their company. They decided, however, to make a "good faith effort" as required by state regulations. After a review of their March 1994 baseline survey, they began to think ECO might work and have positive results for the organization. The company's subsequent compressed work week plans made the front page of the local newspaper. Its ECO plan included fourteen of the twenty-one components described in Figure 6.5. It implemented all fourteen and planned to continue with all fourteen even without a mandatory policy. The company used survey information to get better bus service to the site. To implement compressed work weeks, it reviewed all its policies on scheduling work and the number of hours it allowed an employee to work in a day and made several changes. A year after implementing its ECO program, an employee survey found that 13 percent of its employees were participating in carpools (only a few used transit), 27 percent had chosen compressed work weeks, 8 percent were telecommuting, and 8 percent had moved their start time out of rush hours. Employee participation in ECO programs increased the company's AVO from 1.087 to 1.275 in about one year, over half the increase needed to achieve its target AVO of 1.380. One-third of the increase in AVO was due to changes in automobile use, and two-thirds were due to changes in employees' schedules. When asked about schedule changes, employees overwhelmingly supported the flexibility. Supervisors were also more positive than negative toward flexible schedules:

- 48 percent of employees felt their productivity increased, 44 percent felt it had not changed, and none thought their productivity had declined
- 21 percent of supervisors felt their personal productivity increased, 49 percent felt it remained unchanged, and 9 percent felt it decreased
- 19 percent of staff and 23 percent of supervisors retaining regular schedules said departmental productivity declined
- 89 percent of staff and 59 percent of supervisors choosing alternative schedules said ECO made the company a better place to work
- 43 percent retaining traditional schedules said ECO made the company a better place to work
- Advantages for employees: fewer interruptions, more time for personal business and family, less driving, and longer days, which are more productive
- Advantages for supervisors: improved staff morale and improved teamwork
- Problems for participants: coordinating work and attending meetings
- Problems for supervisors: ensuring proper coverage, scheduling meetings, and coordinating work among staff

Company C, a manufacturing firm, had three sites in Delaware. The largest site had 500 employees. It had recently reorganized and cut back from three to two shifts per day without reducing the workforce. This resulted in larger numbers of employees driving during the morning rush hours. Company C developed different plans for each of its sites, but each emphasized carpooling, biking, and mass transit with a guaranteed ride home. ECO kickoff meetings that gave incentives to attendees drew 90 percent of employees. One site conducted "Wednesday Is Commute Day" and gave free lunch passes to one-fourth of the employees who reported carpooling on Wednesdays. They conducted "Silver Dollar Days" on three unannounced occasions and discovered only one-twelfth of the employees had carpooled those days, about the proportion who carpooled before the program began. After spending between $15,000 and $20,000 promoting carpooling, the company stopped the incentives due to (1) changes in regulations, (2) cost, (3) no state incentive to continue the program, and (4) no apparent lasting impact on commuting behavior. The company planned ten of the components identified in our study, implemented five, and planned to continue two. The company's AVO changed little during the year.

Company D was an educational institution in New Jersey. The initial survey made many employees afraid that the institution was going to start charging for parking for solo commuters or implement other negative sanctions. The ECO program did initially give preferential parking to two-person carpools, which was later increased to three people. The employer's parking unit maintained the database on carpools, and officers checked to make sure that only one car in a carpool was in the parking lot on any single day. About fifty-five people took advantage of a separate program to reimburse transit costs while it operated. The transportation coordinator believed these people were all prior transit riders who viewed transit cost reimbursement as a new employee benefit. The company suspended this component when the state requirements changed. Three people expressed interest in an alternative fuel program that was not implemented because no feasible supply existed. The plan included vanpooling and a guaranteed ride home, but neither component was ever requested. The company also put on transportation fairs, had ZIP code gatherings, carpool parties, transit days, and brown bag lunches. Brown bag lunches included discussion on topics such as the effect of carpooling on insurance. The company's AVO increased from 1.08 to 1.25 in the first year, slightly over halfway to its target AVO of 1.38.

Company E was a food-processing plant in New Jersey with 130 employees. It started its ECO program only because of state regulations, changing the start time of its first shift from 6:00 A.M. to 4:00–5:00 A.M. The employees liked the change, and all new people were hired to start at either 5:00 A.M. or 2:00 P.M. Many employees lived close by and walked or rode bicycles, and the company spent $5,000 to install new bicycle racks and lockers. No public transit served the site. Since they had limited parking, they gave preference to carpools. The company achieved its target AVO during the first year mainly by changing its work schedule.

Conclusion

Large employers in regions with high levels of ground-level ozone were the direct targets of ECO, but their employees were the ultimate targets. Some companies successfully implemented ECO. The Cross County Connections, a TMA in New Jersey, had eighty-five member companies affected by ECO. Most developed plans. Data from twenty-three of these companies showed a combined 10 percent increase from 1.07 AVO to 1.18 AVO in one year, toward a target of 1.38 AVO in two years.

Most companies, however, did not implement ECO. The strongest motivating factor for company compliance with ECO was state regulation. When companies believed their states were serious about implementing ECO, they complied with the regulations. Some would comply due to the normative belief that they should obey the law, and some because they calculated that complying would be better than being fined. Companies would rather pay the cost of developing ECO plans, costs over which they had control, than to pay the unknown cost of penalties over which they had little control.

Cost is also a restraining force, particularly the cost of planning. Companies had to invest personnel time to learn about ECO regulations and how to implement ECO programs. This cost was fairly constant whatever the company size and was, therefore, relatively greater for small companies than for large companies. The more that states and TMAs reduced planning costs by providing detailed information, assistance, advice, forms, and procedures, the more likely companies were to comply. Planning costs appear to be as great or greater under a voluntary program than under a mandatory program, and few companies chose to incur these costs under a voluntary program. The motivating factor of potential fines appears necessary to overcome the initially high restraining factor of planning cost.

The cost to implement plans may be less than the costs to develop those plans. Many companies planning programs under mandatory policy said they would continue to implement them even if the policy became voluntary. The support some ECO programs gained from employees was a motivating factor stronger than the restraining factor of program cost. Good relations with employees can replace state regulation as a key motivating factor in complying with social legislation, once the initial planning is complete.

Most of the ECO programs that companies implemented involved measures over which companies had control rather than measures dependent on external system factors. Work schedules are central to companies' operations, and changing them is the easiest way to meet ECO requirements. Once companies implemented flexible start and stop times, compressed work week options, and opportunities to work at home, they could not rescind them even if they wanted to stop. When companies dropped ECO components, they were most likely to drop those outside their control, such as transit availability.

Companies are very interested and concerned about their workforce, a factor that can either motivate or restrain their compliance with social legislation. A few companies initially held restraining normative beliefs that implementing ECO would interfere with their employees' rights. More frequently, companies did not know beforehand how their employees would view ECO programs. Therefore, they usually planned programs with incentives rather than penalties and focused on the timing of the commute rather than the means of commuting. Most would not have done even this without the motivating factor of government requirements and potential fines. Once planning costs had been incurred, policy implementation could continue under a voluntary program if it did not cost too much and substantial numbers of employees liked the programs.

Most employees commute alone to work every day in their own automobile, and "the journey to work is a commodity," with the choice of route, mode, and time based on transportation costs and benefits (Beaton 1991, 1). External system factors influence employees' commuting behavior and restrain what employers can do in attempting to comply with social legislation. The strongest predictor of solo commuting among employees is the availability of cars in their households. If an individual already has a car available, driving it to work is convenient, saves time, has little direct out-of-pocket expense, and gives the person control over part of the commuting experience. Employers cannot take cars away from employees.

The second strong predictor of solo commuting reflects public opinion against sharing rides, although opinions about ride sharing may be subject to employer influence. Employees who have a positive attitude toward carpooling are less likely to drive alone to work, and employees who do things with other coworkers outside work are more favorable to carpooling than are employees who do not socialize with coworkers outside the job.

Increasing environmental sensitivity can increase favorable carpool attitudes. However, environmental sensitivity had no direct impact on driving alone, and more environmentally positive actions related to more, rather than less, driving alone. The quality of life advocacy coalition had assumed that public support for the environment would help the implementation of ECO. The opposite seemed to be the case.

Company promotion of carpools might affect some people. However, our study suggests that less than one-third of the change envisioned by ECO could be accomplished through carpool programs, especially if company efforts were not reinforced by areawide efforts to change public opinion to be more favorable to sharing rides.

Company promotion of public transportation would probably produce negligible change. Most employees in our study do not have the option of taking public transportation either because they live too far away from bus and train routes or because the work site is too far away from bus and train routes. The local governments who control public transportation were seldom involved in ECO, although legislation did allow them to use federal transportation funds to promote ECO implementation. The few employees who wanted companies to implement ECO transit programs are those already using public transportation. They are generally young, low-income, and female and don't own cars.

The most preferred and most frequently used way for companies to implement ECO was through changes in employee work schedules. Starting the morning shift one-half to one hour earlier immediately produced the policy output of increased AVO as defined in ECO guidance. Required policy outputs could also be achieved by changing from five eight-hour work days per week to a compressed work week of four ten-hour work days, or by allowing employees to telecommute from home on some days. Companies know about scheduling work, and it is something they can control. Most employees like flexible scheduling, but it particularly attracts young, female, upper-income workers with high ideals about teamwork. Once work schedule flexibility is in place, continuing it is likely, no matter what happens with ECO. Unfortunately, employees like flexible work scheduling for reasons other than clean air. How much impact these schedule changes would have on ground-level ozone depends on how cars are used during the time freed from work. Most of the traffic on the road during rush hour is not due to employees commuting to work but to employees already working (e.g., delivering products) or individuals doing errands (Pisarski 1996).

The automobility of the United States is well entrenched. ECO sought to challenge that automobility in one part of life, the commute to work. This policy mandated that employers use their creativity to develop programs to get employees out of their cars during rush hour, when road congestion is at its worst and automobile exhaust contributes most to ground-level ozone pollution. Our data suggest that the quality of life advocacy coalition underestimated the difficulties of implementing ECO when it formulated the policy. Most companies that complied with the social legislation did so by changing their employees' work schedules. However, many companies did not even plan to comply with ECO regulations. They were unwilling to incur planning costs when they believed they could not or should not even try to change employees' commuting behavior. The primary factor motivating them to implement ECO was a state mandate that threatened greater cost if they did not. As we near the end of the policy cycle, policy evaluation takes on great importance as the activities of all the actors involved in implementation are analyzed and the effects of the policy on the problem are examined.

Notes

1. EPA's ECO guidance did not specifically address the issue of requiring employees to change commuting habits. Implicit in the legislation is that somehow companies have to get employees to drive less. However, in an exchange of letters from Senators Joseph Lieberman, Frank Lautenberg, and Harris Wofford to EPA Administrator Carol Browner, the issue was squarely addressed. They specifically

noted that "there is nothing in the Clean Air Act which would force commuters to change their commuting patterns." (See Lieberman, Lautenberg, and Wofford 1994.) Browner's response to the senators was clear, "There is nothing in the Clean Air Act that would force an employee to change commuting habits. An employee may accept or reject an employer's incentives to stop driving alone to work. The Clean Air Act gives employers flexibility to use mass transit, vanpools, carpools, ridesharing, telecommuting, bicycling and walking, or working at home. Many employees will benefit from the ECO program" (USEPA 1994, 1).

2. The 3M Corporation, a twenty-year veteran of travel demand management, provides employees with on-site child care, a cafeteria, and a gym. In 1988 the company went beyond flextime and introduced "personalized work schedules." In addition to the usual carpool and vanpool options, 3M developed a subscription bus program. This program, developed in concert with the local transit company, created special transit services to serve concentrations of employees who lived some distance from the company. See Kuzmyak (1992).

3. Several states, including Maryland, specified later dates in their regulations. This was due in part to the fact that the states issued final regulations later than permitted by the law. Moreover, several states, including New Jersey noted in their regulations that if a company failed to meet its expected increase in AVO by November 1996, it would have one additional year, until November 1997, to demonstrate compliance.

4. In accordance with standard operating procedures, EPA sent all eight states a letter in November 1992 that started the eighteen-month sanctions clock. States then had until July 15, 1994, before sanctions would be imposed. All of the affected states submitted final regulations prior to 1994, although several states later suspended them.

5. Companies have found a way around the Fair Labor Standards Act by changing the company's work week to begin at noon on Friday and end at 11:59 A.M. on the following Friday. This permits the afternoon hours worked during the first Friday to be attributed to the second week when the employee takes Friday off.

6. "Carpool programs" include guaranteed rides home, company-provided vans, carpool locators, and preferential parking for carpools. "Transit programs" include shuttle from transit stops, transit information, subsidized transit fares, shuttle to shopping, rewards for employees using public transit or carpools, on-site daycare facilities, and high occupancy vehicle lanes. "Schedule changes" include flexible hours, compressed work week, and telecommuting.

7. "Negative sanctions" included moving the company nearer transit, allowing carpool parking only, reducing the amount of parking, only hiring employees who work nearby, fining people who drive alone, and making employees move closer to the work site.

7

Policy Evaluation:
Deliberate or Rushed Judgment?

Although policy evaluation comes at the end of the policymaking cycle, it can also be viewed as a basis for new policies. It has the "potential to reframe an issue once thought to be resolved by policy makers" (Gerston 1997, 120). Policy evaluation is the examination and analysis of the activities and effects of federal, state, and local government policies formulated to solve public problems. Sometimes the focus is on the policy itself; sometimes the process by which agencies and organizations implement the policy is under scrutiny, and other times the outputs and impacts of the policy are assessed. Policy outputs are the actions taken by those responsible for implementing the policy, whereas policy impacts are the effects of those actions on the problem.

Policy evaluation can be either formal or informal. Sometimes formal evaluations are required by law. At other times, independent policy analysts conduct formal evaluations in the role of policy brokers. Advocacy coalitions are always informally evaluating policies based on their belief systems. Whoever conducts the evaluation, by whatever means, and for whatever purposes, the object of policy evaluation is "to judge the merits of government programs" (Jones 1997, 175).

Policy evaluation is often viewed as a purely objective activity. It seems that it should be a simple matter of comparing the expectations of the policy as formulated by legislators to the outcomes of the policy as implemented by bureaucrats and private citizens. Rarely is policy evaluation this cut and dried. It is not a value-free endeavor but fits squarely within the political process. In fact, we have defined what most policy analysts call "political considerations" as "informal evaluations" because we believe that evaluations, especially informal ones, determine the winners and losers in the political system, and who wins and who loses is at the heart of politics. Evaluation provides the basis for maintaining or expanding some programs and changing or abandoning others. Further, the sponsorship and purposes of an evaluation may determine its results, even before it is conducted. Changes in system governing coalitions and public opinion can also affect the conduct and results of policy evaluations independent of any changes in the problem.

Like the rest of the policy cycle, policy evaluation is affected by advocacy coalitions and external system factors. Advocacy coalitions often begin evaluating and analyzing policies even before the policy leaves the halls of Congress. Coalitions supportive of the policy will sing its praises, whereas coalitions opposed to the policy will lament its shortcomings. Most of the time, the evaluations done by advocacy coalitions are informal, ad-hoc, and often biased.

In this chapter, we first give a brief history of policy evaluation in the federal government. Second, we review the requirements for systematic formal evaluation. The third section discusses informal evaluations often seen in the media. The final section presents our formal evaluation of the employee commute options (ECO) policy and the modifications we made as a result of the informal evaluations that led to the reformulation of ECO.

The History of Policy Evaluation

Like the other stages in the policymaking process, the meaning of policy evaluation has changed over time. Although government programs have always been informally evaluated, they were not evaluated in a formal or systematic sense before the 1960s. Policy evaluation research was not even a profession before then. Interest in determining program effectiveness increased in the wake of the Great Society programs. These programs provided fertile ground for the beginning of formal evaluation research.

In the early 1960s, program budgeting techniques were introduced into the Defense Department by Robert McNamara, a former chairman of the Ford Motor Company. When he arrived in Washington, D.C., he brought with him the business decisionmaking techniques of operations research. McNamara wanted department heads to think about their programs' concrete objectives and the resources needed to accomplish them. This technique, called program, planning, and budgeting systems (PPBS), required administrators to show what their programs were supposed to achieve and how many of these program objectives would be achievable at various levels of resources. Although attempts to extend PPBS beyond the Department of Defense did not succeed, legislators and administrators accepted the idea of making policy decisions more rational (Starling 1998).

In the late 1960s, Donald Campbell (1969) wrote an article entitled "Reforms as Experiments." He argued that government programs ought to be treated as "natural experiments" and tested to see if they solved the defined problem. He recommended using techniques similar to those used by psychologists in experiments on students and mice in university laboratories. However, government programs do not take place under the controlled settings of a university laboratory. Instead, they take place in the real world, where events and people complicate the situation, making it difficult to isolate the effects of the policy from other factors. To control for the effects of these other factors, Campbell helped create a new method for studying government programs called "quasi-experimental design" (Dunn 1994).

Evaluation research took off rapidly in the early 1970s. The Evaluation Research Society was formed in 1971, new books appeared, and the federal government began to require that some program funds be devoted to policy evaluation. At the end of the 1960s, the federal government spent about $24 million per year on policy and program evaluation. By 1977 this sum had increased to more than $243 million per year (Palumbo 1988).

Contemporary policy evaluation is now an important instrument in the tool chest for judging the value of what governments do. Policy evaluation is a check on the merits of a policy as formulated and its subsequent implementation. It is one of the few tools for deciding if budgets have been spent as intended, rules have been applied as written, and individuals have carried out their assignments as directed. Evaluation not only allows analysts to compare both the process and output of the policymaking process, but it also offers them the opportunity to assess the impact of what has or has not been put into place. Without policy evaluation, both the credibility of the policy and the accountability of those individuals and institutions responsible for its development remain uncertain.

Most policy evaluation textbooks discuss only formal evaluation (Rossi and Freeman 1989). However, informal assessments are frequently more influential in determining the fate of a policy than carefully crafted and rigorously executed formal evaluations. From a "pork barrel" perspective, policies "are judged effective as long as they serve powerful interests" (Patton 1997, 26). Formal evaluations are conducted according to certain accepted principles and formats discussed in the next section. Informal evaluations are recognizable mainly in hindsight. Because our formal evaluation of ECO took place in a turbulent policy environment, we broadened our analysis of the policymaking process to include informal evaluations (i.e., political considerations). The result was our policy cycle advocacy system model and this book.

Formal Evaluation

The goal of formal policy evaluation is to measure systematically what happened and how it affected the problem. Public policies may be formally evaluated as to their processes, their outputs, and their impact. Processes refer to the means by which a policy is carried out. Outputs are the things target populations do in response to policy. Impact describes the consequences for society of government policy, both intended and unintended.

Formal evaluations include several distinct steps. The first step is to define the goals of the policy: What was the problem? What did the policymakers want the policy to do to solve the problem? Second, the analyst must determine the process by which these goals were to be achieved and the problem solved. The third step looks for the theory linking the process to the desired outcomes. Usually, policies specify intermediate outputs that the policymakers expect will achieve the ultimate policy goals. For example, the ultimate goal of ECO was to reduce urban ground-level ozone. The intermediate policy output was a reduction in solo commuting by employees during rush hour. Other intermediate policy outputs included state ECO regulations and employer commute reduction programs.

Fourth, the theory of how processes, outputs, and impacts are related must be identified and made operational. Each concept must be measured. The relationship between the measurements must be specified in hypotheses that can be disproved. It is only at this point that formal evaluation can move to the fifth step and observe the implementation of the policy, gather data, and compare the observed data with the hypotheses. Judging whether a policy is working depends on what aspect of implementation is being analyzed. A program may run smoothly but not produce the desired outputs. The desired outputs may be attained but not affect the problem. If the data are consistent with all the hypotheses, then the policy is judged a success.[1] If the data are inconsistent with all the hypotheses, then the policy is judged a failure. If some hypotheses are supported and others are not, then the policy is judged to need modification or further analysis.

Evaluation Design

Many different types of evaluation designs exist, but most involve variations of three generic types: experimental, quasi-experimental, and nonexperimental. The classic design for evaluation research is the experimental model that uses an experimental group and a control group. Units of the target population (people, companies, states) are randomly chosen to be either in the group that gets the program (experimental group) or in the group that does not get the program (comparison group). Data are collected from both groups before the program begins, during the program, and at the end of the program to learn whether it had any measurable effect on the experimental group. The essential requirement for a true experiment is the randomized assignment of the units of analysis into the two groups (Weiss 1972). Although this design is deceptively simple, allowing only some target units to be affected by the program is often not legal, ethical, or practical.

Additionally, it is difficult in the real world to ensure that no factor other than the experimental program affects the two groups differently, such as the passing of knowledge from the experimental group to the comparison group (Wholey, Hatry, and Newcomer 1994, 159). Although experimental design continues to be the ideal, other designs are more often used in policy evaluation and other social science applications.

Quasi-experimental designs, sometimes called modified natural experiments, also use experimental and control groups to measure the extent to which program activities may have caused observed results. However, the experimental and control groups are not randomly selected. The experimental group is the target population affected by the policy. The control group is selected to be as similar as possible to the experimental group, but it is not subject to the policy being evaluated. Sometimes the control group is in another jurisdiction, sometimes it is in the same jurisdiction but is not affected by the policy, and sometimes it is the experimental group before policy implementation. Some of the best quasi-experimental designs involve single or multiple interrupted time series. These designs look at the trends over time before the program was implemented and compare them to the trends over time after the program. Before-and-after studies are one type of single interrupted time series design. These kinds of studies are good if the program is put into effect abruptly and the effects are expected to occur a short time after the program is implemented.

Nonexperimental evaluation designs do not use control or comparison groups. One common nonexperimental design involves collecting retrospective information from the target population and directly asking them how the program affected them. Another nonexperimental design compares the impacts on those in the target population more involved in the program with the impacts on those less involved in the program. A third nonexperimental design is to conduct case studies of the target population and observe whether the outputs are consistent with the program objectives. The inherent weaknesses of these designs is that they fail to control for alternative explanations. An observed "outcome" may reflect only characteristics of the population rather than a change that occurred as a result of the program. Alternatively, a change in the population may be caused by something other than the program. Although results from nonexperimental evaluation designs are not readily generalized to other programs, they can provide insights into the program and may suggest ways to improve its operations (Patton 1997).

Sources of Information

Sources of information for a formal evaluation have to be considered after the targets are identified. People and organizations can provide information about themselves if they are the targets of policy, or information may be collected about them from others. Organizations cannot respond about themselves if they are the targets, so all information must come from people who know the organization, generally those who are part of it. A third source of information is direct observation by the evaluators.

Information may be quantitative or qualitative, whatever the target units or the source of the information. Quantitative information lends itself to mathematical summarization and statistical analysis. It can be easily compared with other information that uses the same measurements and is especially useful when there are many target units. By its nature, however, quantitative information eliminates detail and may be inappropriate for some analyses. Qualitative data capture a much wider range of detail and are especially valuable for analyzing small numbers of target units. Qualitative data are harder to summarize and do not lend themselves to testing hypotheses. However, they are sometimes easier to discuss with those interested in the evaluation since it is closer to what they can observe for themselves.

Time Frame

Implementing policy takes time, and so does policy evaluation. Ideally, policy implementation and policy evaluation will cover the same time period, but frequently this is not the case. Evaluation often begins well after the start of policy implementation and is expected to provide answers about the policy before the impact of the policy can be readily seen. The policy implementation schedule affects the type of information that can be collected. The best evaluations collect information before, during, and after policy implementation. A series of measures of the target units should be taken as the program unfolds. When it is clear which measurements came first and which came later, the direction of cause-and-effect relationships can be clearly tested. When the evaluation is long enough, it can analyze long-term as well as short-term effects of the policy.

When evaluations have a shorter time frame than the policy implementation, they must compensate in one of several ways. Evaluations can use historical information collected for other purposes. Secondary data may not provide all the information that an evaluation would like, but they can provide some information about the time before the policy was implemented. Historical accounts cannot be influenced by either the policy or the evaluation because they are already complete and are not being made by a person with a stake in the outcome of the policy. Information on past events may also be collected by asking people involved in the evaluation to recall them. Although retrospective information (memory) can directly address issues in the evaluation, it can be influenced by the policy and the evaluation. Further, memory fades with time, and memory of events may be biased by the person's experience with or attitudes about the policy. For example, advocacy coalitions have an interest in what evaluations report and may influence how individuals report past events.

Evaluations may end before the long-term effects of the policy can be seen, due either to pressure to complete the evaluation quickly or to changes in policy implementation that make the evaluation design no longer applicable. Our evaluation of ECO was initially limited by time restrictions on the money to pay for it. The greatest challenge in our evaluation, however, came from changes in the policy itself during our evaluation. Such changes resulting from informal evaluations can wreak havoc with even the best research design.

Informal Evaluation

Informal evaluations are widespread and often begin during the policy's formulation. They increase during the implementation stage, after the first details begin to unfold and the groups targeted by the policy learn the potential costs for them, both in dollars and behavioral change. According to Carol Weiss, "How well a program is doing may be less important than the position of congressional committee chairmen, the political clout of its supporters, or other demands on the budget" (1972, 17). Since informal evaluations have no particular methodological requirements, they span the spectrum from purely anecdotal to more systematic efforts to understand if the program is producing the desired results.

The media play an important role in informal evaluation. Walter Lippman called the media's influence on public policy the "beam of a search light that moves restlessly about, bringing one episode and then another out of the darkness into vision" (1938, 364). The searchlight lands on a policy issue for many reasons, but especially because of the efforts of policymakers, target groups, and advocacy coalitions to illuminate the issue for their own purposes. Whether the media's role is proactive or reactive largely depends on the policy. At times, the media take the lead, seeking to illuminate problems that have reached a critical stage and affect large segments of the population. At other times the media is reactive, waiting to be contacted by individuals or interest groups. Interest groups hoping to get media attention often issue press releases and hold news conferences. The media can help mobilize support or opposition once they get involved and begin focusing on a policy.

ECO and the Media

With ECO, the media played an important but largely reactive role. Few articles about ECO appeared before the implementation stage. Although ECO implementation spanned a five-year period from 1991 to 1995, the media focused on ECO only after significant state and company opposition emerged. Moreover, the media's focus was regional and largely affected by the activities of the economic development coalition in a particular state. When this coalition was particularly active, as in the Chicago air quality region, media attention was high.

We initially began to read newspapers to obtain general information on how the ECO program was being implemented and what companies were likely to know about it and to identify the agency actors at this stage of the policy cycle. As we began to see opposition to ECO grow, we switched to a more systematic analysis of how the media was presenting information on state ECO programs.

We examined the media coverage in detail in four newspapers: the *Baltimore Sun,* the *Philadelphia Inquirer,* the *Chicago Tribune,* and the *Wall Street Journal.* The first three dominate the media markets of their respective air quality regions. Business and economic concerns often make front-page stories if local jobs and businesses are affected. Our primary research focused on Philadelphia and Baltimore, but we decided to include Chicago because the nationwide anti-ECO movement centered there. We included the *Wall Street Journal* in our analysis because it is a national newspaper and a mouthpiece for the economic development advocacy coalition. Further, the *Wall Street Journal*'s influence extends beyond its subscribers because articles are often reprinted in the business sections of regional papers.

We used computerized databases to look at the number and types of articles in the four papers, beginning with January 1991 and ending on December 31, 1996. We used three time periods approximately bounded by four key events in the implementation process: passage of the Clean Air Act Amendments of 1990 (CAAA-90), issuance of the Environmental Protection Agency's (EPA's) guidance, suspension of ECO implementation in Pennsylvania, and congressional change to ECO to make it voluntary. Articles were identified and classified into one of three categories by their overall orientation—pro-ECO, neutral or informative, and anti-ECO. Our review of newspaper articles suggests the media became increasingly active as the focal point of ECO activity moved down from Congress to the states and to the companies. As business organizations got the word out to their members, employers started to oppose ECO, and it was their opposition that fueled media attention. The number of articles increased and changed from initially neutral and even pro-ECO to anti-ECO. (See Table 7.1.)

Guideline Development: January 1991–December 1992

The first period yielded few articles about ECO, and most of them were informative rather than evaluative. Of the two articles that appeared in the

Table 7.1 Number of ECO Articles in Four Newspapers by Year and Content

	1991–1992 Guideline Development			1993–1994 State Implementation			1995–1996 Company Implementation		
	Pro	Neutral	Anti	Pro	Neutral	Anti	Pro	Neutral	Anti
Wall Street Journal	—	1	1	1	3	7	2	4	2
Baltimore Sun	2	7	2	2	7	12	6	3	9
Philadelphia Inquirer	—	4	5	4	1	13	2	—	10
Chicago Tribune	—	3	—	7	8	27	3	2	19
Total	2	15	8	14	19	59	13	9	40

Wall Street Journal, the first was a general overview of ECO and focused on the implementation schedule. The second article was clearly anti–government regulation and simply listed ECO as one of the onerous and costly provisions in the CAAA-90.

The number of articles that appeared in the regional papers varied from three in the *Chicago Tribune* to eleven in the *Baltimore Sun.* The *Baltimore Sun* was the only paper during this period to publish articles with informal evaluations supporting ECO. Two articles (Sachs 1992; Morris 1992) emphasized the benefits of carpooling for employees (saving money) and for urban areas (cleaner air). Most of the *Baltimore Sun* articles during this period were neutral, although two articles provided informal negative evaluations. These anti-ECO articles questioned the authority of the CAAA-90 to require employers to get their employees into carpools. One article suggested that the "proposed rules to battle air pollution could, in extreme cases, give Baltimore-area employers the right to fire workers who refused to take such steps as car pooling or working at home" (Mullaney 1992, 12C).

The articles in the *Philadelphia Inquirer* and *Chicago Tribune* followed a pattern similar to those in the *Baltimore Sun.* All of the seven neutral articles focused on the proposed regulations and included general information about the Clean Air Act. The five anti-ECO articles all appeared in the *Philadelphia Inquirer.* One article focused on the issue of employer liability in promoting car- and vanpools (Stranahan 1992). Another article speculated on the cost of the program and its inconvenience for some employees (Struzzi 1992). No anti-ECO articles appeared in the *Chicago Tribune* during this period.

Overall, the tone of the articles in all four papers during this period was informative and descriptive rather than adversarial. The media were doing their job of passing on information to their readers. Articles most often quoted state officials who limited their remarks to factual information about the law, the schedule, or the times and places of public hearings. Few employers were quoted.

State Implementation: January 1993–December 1994

By 1994 most affected employers had heard about ECO and were engaged in developing their company plans or looking for ways to avoid the regulations. In some regions, local business groups had already taken anti-ECO positions; in other regions, the opposition was still weak. Almost four times as many articles appeared during the two years of state implementation as during the two years of guideline development. The tone of the articles became more adversarial, and employers were quoted more often than state officials.

The *Wall Street Journal* published most of its articles on ECO during this period. Seven of the eleven articles were anti-ECO, and two of these were reprinted in both the *Baltimore Sun* and the *Philadelphia Inquirer.* The first article was an editorial, pointedly entitled "Hitting the Carpool Brakes" (*Wall Street Journal* 1994b, A10). The article focused on the Los Angeles region and noted that regulators there were considering taking a "step back from an economic chasm" and revising their employer-mandated carpooling regulations. It went on to recommend that other regions take note of this development. The second article was a feature piece on the front page entitled "Head-on Collision: Cut Auto Commuting? Firms and Employees Gag at Clean-Air Plan." The article emphasized the limited gains and mass inconvenience of ECO and concluded that it was a bureaucratic nightmare (Caleb 1994, A1).

The articles that appeared in the three regional papers were even more adversarial and combative than those in the *Wall Street Journal.* The *Chicago Tribune* published forty-two articles during this period, and twenty-seven of these emphasized the growing anti-ECO sentiment in the region. The *Chicago Tribune*'s extensive coverage of ECO was influenced in part by the active anti-ECO role played by Governor Jim Edgar, Congressman Donald Manzullo, and the Chicagoland Chamber of Commerce. More than half of the articles appeared during a six-month period between November 1993 and April 1994. Businesses lobbied intensely to delay the program during this period since Illinois regulations required large employers to begin registering in February 1994. The titles of several articles suggest negative informal evaluations of ECO—"Commute Mandates Under Fire" (Ibata 1994a); "Local Firms Buck U.S. Law as Clean Air Deadline Nears" (Ibata 1994c); and "Foes of Clean-Air Act in Overdrive: Business Groups Fear High Costs Will Outweigh Benefits" (Ibata 1994b). However, seven pro-ECO articles also appeared. Most appeared under letters to the editor and reflected the struggle between the economic development and the quality of life advocacy coalitions. One article written by the executive director of the American Lung Association of Metropolitan Chicago emphasized the benefits of the program and took the local business community to task for their exaggerated cost predictions (Kirkwood 1994).

Articles in the *Baltimore Sun* and the *Philadelphia Inquirer* also reflected growing business opposition. Most of the neutral articles appeared in 1993, whereas the anti-ECO ones appeared in 1994. Both states shared similar experiences during this time. Business opposition succeeded in convincing the governors of both states to suspend the regulations, and they applied pressure on legislators not to reverse the suspension. One article published in the *Baltimore Sun* reported that "state officials make it plain that they don't want to impose the controversial commuting program that has drawn complaints from Baltimore area employers..." (Wheeler 1994a, B1). The same article quotes a spokesperson for the American Lung Association as saying, "Hey, there's bigger things to fight about" (Wheeler 1994, B1). This attitude, that there were more important issues to fight, suggested that the quality of life advocacy coalition had given up on ECO. With no strong coalition to promote the program, state officials could easily side with employers and the economic development advocacy coalition. By 1994 the media reported the view that everyone was opposed to the regulations.

Company Implementation: January 1995–December 1996

By 1995 business opposition to ECO was nationwide. Cost was an issue for many businesses. However, many employers informally evaluated ECO negatively based on their belief that government should not tell businesses and individuals what to do and when to do it, one of the basic beliefs of the economic development advocacy coalition. In the Chicago region, the state's four most powerful business lobbies backed a joint resolution that was unanimously approved by the Illinois General Assembly. The resolution demanded that the U.S. Congress do away with ECO and mandatory trip-reduction programs (Christian and Crimmins 1995). Four states suspended their ECO regulations during the first six months of 1995: Pennsylvania, Illinois, Maryland, and Texas. Most of the remaining states made their programs more flexible.

Mary Nichols, EPA assistant administrator for air and radiation, was quoted as saying, "The air emission reductions from these programs are minuscule, so there's not any reason for the EPA to be forcing people to do them from an air quality perspective" (Crimmins and Kendall 1995, A1). Although EPA's "new" policy had not been formalized in writing, it did not seem to matter to the media. All four newspapers wasted no time in reporting that EPA did not intend to enforce the regulations. The obvious contradiction of EPA mandating regulations that it was not going to enforce was a perfect story. All four papers carried the initial story, especially the *Chicago Tribune,* which repeated the quotation in many subsequent anti-ECO stories.

During this period, the anti-ECO articles in the *Chicago Tribune* outnumbered, by nearly two to one, similar anti-ECO articles in the *Philadelphia Inquirer* or *Baltimore Sun.* This was due in large part to the efforts of both Governor Edgar and Congressman Manzullo. The governor held three news conferences between February and April 1995 to denounce and then suspend ECO. Simultaneously, hearings were being held on Congressman Manzullo's bill (H.R. 325) to make ECO voluntary. The *Chicago Tribune* followed these informal political evaluations very closely. According to Manzullo, "I made it a point to get press releases to the papers. And once the stories came out, I copied them and sent them out to all my bill co-sponsors and anyone else I could think of" (Manzullo 1997). The *Chicago Tribune* articles helped Manzullo's effort.

Once Manzullo's bill passed and ECO became voluntary in December 1995, pro-ECO articles began to appear in all four papers. Most of these articles, however, focused on voluntary efforts that were already in place in some regions. Both Maryland and Illinois had voluntary programs in place for the 1996 summer ozone season. As an optional rather than a required measure, ECO activities were embraced by agencies that were freed from the regulatory obligations that previously burdened them.

A Formal ECO Evaluation Design

We developed our formal evaluation of ECO shortly after Congress passed the CAAA-90. Those who were aware of the ECO provision had little doubt that the policy as formulated would have a significant impact on urban areas, their businesses, and employees. Few on the evaluation team had reason to think that ECO would not be implemented. Given the short schedule for ECO, we believed that a three-year evaluation project (the maximum time for available grants) would yield useful information. The branch of EPA responsible for research grants agreed. However, even before our formal evaluation began, the policy environment began to change and continued to change throughout the project. EPA delayed its guidance, states refused to cooperate, advocacy coalitions protested, target groups complained, and the expected implementation schedule kept being extended. These events required us to modify our original formal evaluation design.

Our Original Evaluation Design

Our evaluation was originally planned as a quasi-experiment, in the Baltimore air quality region. We timed our research to coincide with the legislated timetable of the ECO program. We planned to begin in early 1994 as companies in Maryland started planning for ECO. Our evaluation would end in early 1997, a few months after companies submitted their final compliance reports to the state. We believed that companies and employees in the Baltimore metropolitan area would be representative of companies and employees in the other metropolitan areas of the United States subject to the ECO requirements.

The purpose of our evaluation was to study the economic and social impact of the ECO legislation on businesses and their employees. The original research design had five objectives; the sixth was added later to capture state differences in policy implementation:

1. document the economic effects of the regulations on business;
2. identify the economic and social effects of the regulations on employees and on employer-employee relationships;
3. document the effectiveness of business efforts to change employee commuting behavior;
4. identify how various sources of information and assistance can affect cost and effectiveness of compliance strategies;
5. provide information back to the system to result in more efficient achievement of compliance goals; and
6. document the development of and differences between four state regulations and their effects on employers.

The model for the original evaluation adapted the compliance with social legislation (CSL) model to analyze the factors associated with commute change. (See Figure 6.1.) We intended to apply the model to changes in employees' commuting behavior, from driving alone to nonsolo commuting. In this context, the model would consider the factors within companies that motivated employees to comply or restrained them from complying with social legislation.

The methods we selected would have involved 120 businesses over a three-year period as they developed their plans, implemented their programs, and reported to the state. An "experimental" group (approximately sixty companies) would be recruited from Baltimore-area businesses who used the services of the two transportation management associations or assistance from Towson University. The "control" group would be a matched set of Baltimore area businesses who did not use these services as they implemented ECO programs. Data would come from annual surveys of employees and companies and from the ECO plans and reports filed by the businesses with the state. The expected data are shown in Box 7.1. The evaluation team at Towson University offered to help companies by administering their required employee surveys.

The Modified Evaluation Design

The first change in our design occurred when Maryland first delayed its implementation of ECO. We switched our evaluation focus to the Philadelphia air quality region, where the other three states already had ECO regulations in place. The Philadelphia region was not only convenient for us, but it was the only region in the country that involved four state ECO regulations. The EPA believed that comparing ECO's implementation in several states enhanced our evaluation, and we believed that the four states would be representative of the eleven states affected by ECO.[2]

The change in regional focus complicated the selection of companies. We had to purchase a commercial list of companies that had several limitations for our study: (1) each record on the list represented a company rather

---

Box 7.1   Data Planned for the Evaluation

Employee Survey
- Commute behavior (as required by regulations)
- Cost of commuting
- Attitudes toward commuting
- Environmentally related actions
- Attitudes toward the environment
- Work group cohesion
- Attitudes toward potential employer programs
- Relevant demographics

Organization Surveys
- Economic costs and benefits
- Social costs and benefits
- Sources of ECO information
- Process to develop and modify plans
- Process to implement and manage plan
- Process to monitor and report results
- Subjective assessment of the impact of the regulations

ECO Plans and Reports
- Program components
- Program costs
- Target APO
- Achieved APO

than a site; (2) companies with headquarters outside the Philadelphia area were not always listed; (3) company size was reported as an aggregate of all sites rather than for each individual site; and (4) company statistics, especially number of employees, were not always current. Company mergers, small headquarters staff, and recent downsizing resulted in many sample companies having fewer than 100 employees when we contacted them.[3]

The switch of our evaluation focus from Baltimore to Philadelphia delayed the start of our recruitment of companies. The recruitment process itself took longer than expected. Our study began in February 1994, and we decided in April 1994 to focus on Philadelphia. Our recruitment of companies continued for a year until June 1995. We recruited about two-fifths (112 of 274) of the companies overall, but this was an average of a high recruitment rate at the beginning and a low recruitment rate at the end. (See Figure 6.2.) As states suspended or changed their ECO requirements, company recruitment became harder.

The first data about companies came from notes taken by evaluation team members when they talked with companies to explain the evaluation and request their participation. This information was standardized during the 1996 follow-up survey on the status of company ECO programs. We also collected data on companies in June 1995. Six booklets constituting the organization surveys were sent to different company administrators. Questions on ECO program costs went to transportation coordinators; questions on company processes used to implement change went to human resources executives; questions on the importance of various goals to the company went to line executives. The remaining three booklets were sent to employees at different levels of the companies to provide different perspectives on company culture.

We anticipated that collecting employee data would be easy once companies decided to comply with ECO legislation and participate in our study. We designed an employee survey to collect all the information companies needed to report to their respective states, along with the information we needed for our evaluation design. The survey also collected information we thought would be useful for companies planning ECO programs. We pretested the survey at Towson University and prepared a Spanish version.[4]

We soon encountered several problems in administering our survey. New Jersey prohibited companies from using surveys that changed any words or asked additional questions anywhere except at the end.[5] Delaware required companies to include a set of questions to be used for its state transportation model. Maryland and Pennsylvania accepted our employee survey form but suspended their programs before many companies conducted employee surveys. Overall, our research obtained some kind of baseline employee survey data from thirty companies and follow-up employee survey data from five companies. Table 7.2 shows the characteristics of 4,787 employees from fifteen companies that used our form for their baseline employee surveys.

In our original formal evaluation design, we did not intend to systematically monitor newspaper articles or state regulations. We started informally collecting this information to help recruitment, design the employee survey, and provide a context for later analysis of data. Once we realized that

Table 7.2   Social and Demographic Characteristics of Responding Employees

	Percent	Number
Sex		
Male	39.7	1,701
Female	60.3	2,584
Missing	—	502
Age		
Under 25 years	8.5	361
25-34 years	34.1	1,455
35-44 years	28.9	1,230
45-54 years	19.6	837
55 years and over	8.9	380
Missing	—	524
Education		
Not high school graduate	5.1	211
High school graduate	31.6	1,301
Some college	26.5	1,090
College graduate	26.9	1,107
Advanced degree	10.0	410
Missing	—	668
Type of Job		
Service, maintenance	11.6	472
Operative	9.0	365
Skilled crafts	6.8	277
Secretarial, clerical	23.8	970
Technical	11.6	473
Sales	3.6	145
Professional	21.3	867
Administrator, manager	12.5	509
Missing	—	709
Family Income		
Under $25,000	25.0	910
$25,000 to $49,999	38.6	1,404
$50,000 to $74,999	21.4	778
$75,000 and over	15.0	547
Missing	—	1,148
Total	100.0	4,787

political considerations (informal evaluations) were greatly affecting the development of state regulations and the implementation of ECO, we started systematically collecting, analyzing, and comparing the newspaper articles as well as the drafts and final copies of state regulations.

Part of our planned formal evaluation included giving information gathered from the project to employers through annual conferences. These conferences would provide employers with the opportunity to share information directly with one another and give us further insight into program philosophies, processes, and outcomes. The first planned workshop was canceled because we simply had nothing to report. We canceled the second planned workshop because few companies showed any interest in attending. Enough companies expressed interest during our follow-up survey that we held two workshops during the third year of the evaluation (Center for Suburban and Regional Studies, 1996a and 1996b). These workshops involved twenty-four companies and organizations and included results from both our formal and informal evaluations of ECO.

As our formal evaluation progressed, we also disseminated information through papers and articles to academic audiences in environmental studies (Bonham and Farkas 1995), transportation studies (Farkas 1996), urban affairs (Bonham 1996; Bonham, Burnor, and Marzotto 1994, 1995; Marzotto 1996, 1998), and public administration (Marzotto, Burnor, and Bonham, 1995).

A Comprehensive Evaluation

The impact of informal evaluation changed ECO policy implementation sufficiently that our planned formal evaluation had little meaning by itself. Our formal evaluation focused on the outputs of changed employee commuting. These outputs simply did not exist or could not be measured because few companies implemented ECO programs, and only a few of those that did implement programs conducted postprogram surveys of employee commuting to measure change. Further, no key decisionmakers had any interest in the results, an important criterion for the purpose of evaluations (Patton 1997). Was there no value in the information provided by 112 businesses and almost 5,000 employees? Was there no value in information gleaned while following the implementation of ECO at the federal and state levels? No, not if we kept our original or modified formal approach to evaluation.

We decided to take a broader perspective on policy and to consolidate what we had measured formally and learned informally. We looked at various ways to approach policy analysis and found that, when used alone, none satisfactorily explained what had happened to ECO. We therefore integrated several different approaches to develop the policy cycle advocacy system model.

Conclusion

Formal policy evaluation, with standardized methodologies, developed as a discipline in the 1960s and is frequently thought to occur at the end of the policymaking cycle, after policy implementation is well under way. In order to conduct a formal evaluation of ECO at the beginning of the implementation stage, we approached EPA with a proposal to analyze the economic and social impact of ECO on businesses. This early start on evaluation allowed us to develop a quasi-experimental design using a rigorous approach to sampling, measurement, and theory-based hypotheses. Our formal evaluation assumed that the policy would remain constant once it was formulated. We focused on the final phases of implementation—what happens when policy meets private citizens. We took a bottom-up approach to policy evaluation, ignoring the activities of government and focusing on the actions of the private sector.

Most evaluation, however, is informal and driven by the basic, policy, and instrument beliefs of advocacy coalitions. Informal evaluations begin immediately after the policy is formulated and may not consider whether policy produces the intended outputs and impacts. The value of informal evaluations is measured more often in political rather than scientific terms. During ECO's implementation, changes in variable system factors shifted resources from the quality of life advocacy coalition to the economic development advocacy coalition, and the latter's informal evaluations of ECO began to dominate.

Our analysis of media coverage of ECO tracks this shift from positive to neutral to negative informal evaluations. As the informal evaluations began to increase, the policy began to change. This change constrained our planned formal evaluation, requiring it to change and lose some of its scientific purity. Informal evaluations continued to increase until they reached a level that compelled policymakers to reformulate the policy. The dynamics of policy implementation made a static formal evaluation moot.

However, the process of formal evaluation can focus research attention on policy implementation and provide valuable information for future policy making. Although our formal evaluation did not follow our planned outline, it did collect data documenting ECO implementation; tracing decisions by sovereigns, agencies, and organizations; and projecting the decisions by individuals. In the process, it measured variable system factors that affect the development and implementation of policy within policy subsystems. Most important, though, data collected during formal evaluation can provide a platform from which to view the broader picture of public policy. Our formal evaluation provided us with the basis for developing what we believe to be a better way for understanding the dynamics of government attempts to solve public problems through public policy.

Other policies are unlikely to emerge in exactly the same way as ECO did in the CAAA-90. However, the principles learned from our ECO evaluation and developed in this book help project what might happen with future policies in the clean air, transportation, and urban subsystems. ECO policy has been reformulated as voluntary rather than mandatory based upon informal evaluations. We turn our attention now to what contributed to that reformulation, and the impact a reformulated ECO policy is likely to have on the problem of air pollution from automobiles in urban areas.

Notes

1. Formal research and evaluation never "proves" the hypothesis that the policy works. It technically can only reject the "null" hypothesis that the policy does not work. Scientific proof is accumulated with many rejections of the null hypothesis (that the policy does not work).

2. Although the switch in urban areas eliminated the opportunity to test the effect of transportation management associations on company programs, it provided us with the opportunity to study the effect of different state regulations. As a result of this change, we added the sixth objective to our original list. We also realized that the compliance with social legislation model was more appropriate for understanding company willingness to implement ECO programs than it was for understanding employee willingness to change commuting behavior in response to company ECO programs.

3. The Delaware sample presented special problems. Delaware tax laws attract the "paper" headquarters of many large companies, which are maintained by a few employees. Three-fifths (61 percent) of the Delaware employers identified by Dun and Bradstreet as having 100 or more employees had either only a few employees or no telephone listing at all. The research team made a decision to supplement the Dun and Bradstreet Delaware sample with private-sector members of the New Castle Transportation Management Association. About half (thirty-two) of the association's members were not listed by Dun and Bradstreet. Most of these companies were located in Delaware, but one had its headquarters in the City of Philadelphia and one had its headquarters in New Jersey.

4. Our employee survey was designed to be completed by employees using a regular pencil and then automatically read by machine. Nonreaders or readers of other languages relied on the informal translators they used for other work-related communication.

5. Additional questions could only be added at the end of the New Jersey survey. This meant that the study's employee survey form could not be used as a baseline survey in New Jersey unless companies were willing to give both survey forms to their employees. We did help a number of New Jersey companies collect and tabulate their 1994 New Jersey baseline employee surveys but could not use this data for our analysis. New Jersey companies could use the study's employee survey for a follow-up survey in 1995, and some did so.

8

Policy Reformulation:
The Problem Revisited

We have a responsibility to act quickly to fix federal programs, such as this one, that have proved unworkable.

—Senator John Chafee (R-R.I.)[1]

How do decisionmakers fix a policy that has been informally or formally evaluated as unworkable? Should they continue to implement it? Should they terminate it? Should they change it?

Most policies are not terminated. Once an idea has been formulated into public policy, that idea tends to remain there. However, the policy may be changed or reformulated to reduce shortcomings or satisfy advocacy coalitions opposed to the policy. Changes in the external system environment cause policy to be reformulated: new people come into office, new technologies emerge on the scene, or the problem gets worse. The policymaking process recycles—the problem is redefined, it moves up again on the institutional agenda, it is reformulated, and an apparently new program is implemented. Our model illustrates the continuous nature of the policymaking cycle with arrows (see Figure 1.3) leading from the later stages in the cycle back to the earlier stages. Advocacy coalitions stay involved as the cycle repeats. Early calls for change in the employee commute options (ECO) policy developed steadily into a relentless drumbeat for change. Opposition from the economic development advocacy coalition, noninvolvement by the quality of life advocacy coalition, uncertain and confused implementation by federal agencies, lack of cooperation from the states, and changes in governing coalitions in Congress contributed to a demand for change in ECO. Policymakers, however, did not terminate ECO; instead they reformulated it.

The Environmental Protection Agency (EPA) gave states some flexibility in implementing ECO when it first published guidelines for the program. States took advantage of this flexibility and developed their own individualized regulations based on input from companies, business groups, and other interested groups. This flexibility and the growing pressures from the informal evaluations by states and companies soon found the EPA promoting a confusing and unpopular program. To reduce the growing opposition, EPA tried to increase flexibility and soft-peddle enforcement (i.e., loss of federal highway funds). This approach generated even more opposition as companies questioned the logic of their having to implement a policy that EPA would not enforce. Eventually, Congress reformulated ECO from a mandatory to a voluntary policy.

The two advocacy coalitions had differing influences at different stages of the policy process. The economic development advocacy coalition did not effectively organize opposition until 1992, when the states began writing their ECO regulations during the implementation stage. To get companies and business groups to buy into ECO, states sponsored workshops, forums, and meetings to educate employers. The people attending these ECO workshops and training sessions were company representatives, transportation experts, air quality professionals, and business organizations. Although useful for disseminating information, the workshops provided the opportunity for the economic development coalition to solidify its opposition to ECO by giving its members an opportunity to talk among themselves and compare notes. In general, members of the quality of life advocacy coalition did not attend these workshops and, therefore, were unable to counter the growing discontent among companies and business groups.

Initially, states attempted to change ECO by getting all or part of their air quality region reclassified out of the ECO program. Companies then tried to get state implementing agencies to scale back regulations, include more flexibility, and provide more assistance to companies. Finally, states, companies, and business groups joined forces and asked Congress to change the Clean Air Act itself. The 1994 congressional election, in which the Republican Party won control of Congress, changed the external system factors that affected the balance of power between advocacy coalitions. The Republicans ran on a platform opposing federal unfunded mandates.[2] Once in control of Congress, Republican committee chairs looked for ways to remove the burden of federal mandates from state and local governments. The economic development advocacy coalition found backing for its opposition to ECO in this new, more supportive, congressional atmosphere. ECO was amended in 1995 and became a voluntary program.

Undoubtedly some critics thought that this would be the end of the program, but this has not been the case. The number of regions with ECO programs has grown from the originally mandated ten to more than thirty under the voluntary policy. The reported success of many of these voluntary programs convinced EPA to issue guidance on how states can get credit for ozone reduction resulting from their voluntary ozone reduction programs. This is quite a change for a policy assumed to be unworkable in 1995. How and why did this change occur? We turn now to an analysis of the end of mandatory ECO and the rise of voluntary ECO.

Ending Policy: Congress Reconsiders ECO

The First Attempt at Reformulation: Unsuccessful

On June 16, 1994, Representative Donald Manzullo (R-Ill.) introduced "a bill to amend the Clean Air Act to provide for an optional provision for the reduction of work-related vehicle trips" (H.R. 4589). The bill proposed to amend the ECO provision of the Clean Air Act Amendments of 1990 (CAAA-90) by substituting the word "may" for the word "must," thus converting the statute from a mandatory to an optional employer trip reduction program. Although Manzullo was not a member of the House Energy and Commerce Committee during the 103rd Congress, he became interested in the Clean Air Act and, specifically, the ECO provision after attending a breakfast meeting in his district where he was confronted by angry businesspeople (Manzullo 1997).[3]

The bill was cosponsored by a solid cadre of Republican representatives, all but one from states with severe nonattainment areas. In his introductory remarks, Manzullo stated that "my bill will allow states to decide if they want car pooling to be part of their clean air plan.... My legislation sends a message to EPA that there needs to be more flexibility in the law" (U.S. Congress, House 1994, 4556). The bill was referred to the Committee on Energy and Commerce, which referred it to its Subcommittee on Health and the Environment. However, since John Dingell (D-Mich.) chaired the Committee and Henry Waxman (D-Calif.) chaired the subcommittee, both Democrats who had worked hard for the CAAA-90, neither hearings nor floor actions were scheduled during the 103rd Congress. The bill died at the end of the 103rd Congress, but Manzullo emerged as the point man for the anti-ECO forces.

The Second Attempt at Reformulation: Successful

Manzullo reintroduced his second bill (H.R. 325) to make ECO voluntary on January 4, 1995, the first day of the new legislative calendar. The 104th Congress was very different from the 103rd Congress because Republicans controlled both houses for the first time in forty years. The Republicans operated under the "Contract with America," which included a commitment to reduce unfunded mandates. ECO was an unfunded mandate and a perfect candidate for the new Republican governing coalition to prove their commitment to the "contract." In this dramatically changed political environment in which Republicans replaced Democrats as committee chairs, Manzullo's bill received serious consideration. This change in the governing coalition gave added resources to the economic development advocacy coalition and severely constrained the power of the quality of life advocacy coalition.

This time, the economic development advocacy coalition did not rely on a single House bill. Rick Santorum, the newly elected Republican senator from Pennsylvania, introduced a companion bill in the Senate to make ECO voluntary (S. 328). House Majority Whip Tom Delay (R-Texas) introduced a more extreme bill (H.R. 478) to repeal ECO altogether. Additionally, the Emergency Supplemental Appropriations and Recission Act (H.R. 1158) included a section prohibiting EPA from spending any money on ECO. Thus, the economic development advocacy coalition had three options: (1) it could launch a frontal assault to get rid of ECO altogether, (2) it could work for a reformulation of ECO into a voluntary program, or (3) it could go through the back door and stop ECO's implementation by halting the money. This last option was used at the state level in both Pennsylvania and Maryland.

The coalition focused on the second option, and they began the drive to reformulate ECO. This time the House Committee on Energy and Commerce, no longer chaired by Dingell, assigned the bill to its Subcommittee on Oversight and Investigation. Manzullo tried to get Dingell and Waxman to cosponsor his bill since they were still ranking minority members. However, when Manzullo approached Dingell with the idea, Dingell is reported to have declared, "Not in my life time, nor in my children's or my grandchildren's life time do I intend to open up the Clean Air Act" (Manzullo 1997). It was clear from this remark that the thirteen years Dingell had spent negotiating CAAA-90 was more than enough struggle for one person. His declaration also reflected a common fear expressed by some members of Congress that opening one part of the Clean Air Act to amendments would signal that other parts of the act could be changed as well.

Subcommittee Chair Joe Barton (R-Texas) began two days of hearings on Manzullo's bill (February 9 and March 16, 1995) by noting that:

> Although the House-approved version of the 1990 Act did contain sections on reducing vehicle miles traveled in certain nonattainment areas, the precise language of ETRP [ECO] was established in conference based on section 183(c)(2) of Senate 1630, the Senate bill. I point this out not to blame the Senate but to suggest that, at least in the House, there is less than a full legislative record with respect to this particular program. Regardless of history, though it is now abundantly clear to the 28,000 businesses and the 12 million Americans affected by ETRP.... As we

should hear today, ETRP in theory and ETRP in reality may be two different things. (U.S. Congress, House 1995, 157)

Despite his protestations to the contrary, Barton was seeking to distance the House from the original ECO provision and place responsibility for its inclusion in the CAAA-90 squarely on the Senate.

Except for EPA Administrator Carol Browner and her deputy Mary Nichols, who were expected to support the law, only one witness, representing the Association of Commuter Transportation, found anything good to say about ECO as formulated in the CAAA-90. Even those who had something positive to say about their own state's experience implementing ECO ended their comments by suggesting that something needed to be done about the federal program. For example, Ron Wyden (D-Oreg.) said:

> The Oregon ECO program has not only been an effective clean air strategy but it has also yielded some additional dividends for the business community. For example, under the Oregon approach, the Nike Corporation found that it did not have to build 220 additional parking spaces, for a savings of $770,000.

Yet despite these glowing remarks, Wyden went on to say, "the implementation of this program has left much to be desired" (U.S. Congress, House 1995a, 159).

By 1995, California's love affair with commute trip reduction was over. Several witnesses from California confirmed that the cost of their program (Regulation 15) far exceeded the benefits. As Robert A. Wyman, chair of the South Coast Air Quality Management District's Committee to Evaluate the Implementation of Regulation 15, noted:

> We need to do more than simply put pressure from the employer on the employee to create a convenient alternative. If there is no proximity to transit or to other forms of transportation which would be high occupancy in nature, there may be very little that even a well meaning employee can do to react favorably to that pressure. Clearly one of the missing links is adequate transportation alternatives. (U.S. Congress, House 1995a, 204)

Perhaps the testimony that had the greatest influence on the reformulation of ECO came from EPA assistant administrator Mary Nichols, as she tried to support ECO for the quality of life advocacy coalition. Nichols knew she was going to be in the hot seat for her published remarks made a month earlier (January 1995) in which she stated that EPA was not going to enforce ECO. Her statement pleased companies but angered lawmakers who questioned EPA's authority to decide which laws it would or would not enforce. Nichols's appearance gave the subcommittee an opportunity to interrogate her about EPA's seemingly arrogant stance on ECO. In a three-way exchange between Nichols, Chairperson Barton, and Representative James Greenwood (R-Pa.), the frustrations of all three surfaced:

> Mr. Greenwood: What is the legal basis for EPA's position that it can suspend enforcement of the Employer Trip Reduction program?
>
> Ms. Nichols: First of all, EPA enforces the Clean Air Act. We do not suspend enforcement of the Clean Air Act. We have from time to time under very special circumstances indicated that we would exercise prosecutorial discretion in certain situations. . . . We do not intend to enforce against individual employers under the Employer Trip Reduction Program. In fact, the general statutory scheme under the Clean Air Act is one in which states have the primary responsibility for enforcing all measures that are within their SIP's.... Our position on this issue is that EPA must require a state to submit an ECO program. We do not enforce the plan... ; that is the states' responsibility. We have given guidance to them that has given, we believe, a great deal of flexibility in terms of determining whether that plan is enforceable. That is, we have said that a State does not have to have each employer plan reviewed on a plan by plan basis.
>
> Mr. Barton: Are you saying, then that you still expect each employer that employs more than 100 employees to develop such a plan and you expect them to turn it into the state, and you are just not going to enforce sanctions on the state if the state doesn't check to see if the plan is being implemented? Is that what you are saying?
>
> Ms. Nichols: No. What I'm saying is, so far where we are with respect to this provision is states have done their job. They have developed plans. They passed the legislation to implement these programs. They have submitted the plans to us. We have received those plans and we found them to be complete.
>
> Mr. Barton: If I am understanding you correctly, you are saying... you're not going to hold the employer responsible, but you're going to hold the states responsible for this. Is that what you have told me? I am not being argumentative with you. I think it's a key thing. I'm of the belief that this is something that simply cannot be implemented and that we are going to have to change the act.
>
> Ms. Nichols: Mr. Barton, I have a figure here... 22 employers who implemented trip reduction activities and showed that the five companies that had achieved the greatest results reduced the vehicle trips by 35 to 48 percent, and these were not all facilities that were located near transit.
>
> Mr. Barton: The panels that came from all over the country. We just picked imbecilic representatives of industry? We just picked the wrong people?
>
> Ms. Nichols: No, sir, and I don't mean to be disrespectful. . . . I simply wanted to say that we do have counter examples if you had asked to assemble a panel of people from communities where the programs have worked. (U.S. Congress, House 1995, 214–218)

The exchange underscores two elements common in the congressional hearing process. First, as with the original hearings on ECO, the witnesses were selected to support the subcommittee's point of view. Subcommittee members viewed with disdain Nichols's attempt to bring balance to the hearing by sharing some ECO success stories. Second, subcommittee members delight in grilling administrators. The congressional oversight function is inherently adversarial, especially when administrative testimony contradicts a subcommittee's point of view. When members are looking for problems with programs, they expect everyone, even administrators, to support their opinion (Davidson and Oleszek 1990).

ECO remained in limbo until December 1995. On March 15, 1995, before the House subcommittee concluded its hearings on ECO, EPA asked the Clean Air Act Advisory Committee (CAAAC) to look for any program design options available to states under the current law. EPA planned to provide new guidance to states and employers. The committee, a blue-ribbon group of state, industry, and public interest leaders from both the economic development and quality of life advocacy coalitions, acted as a broker. It reported its assessment to EPA on April 21, 1995, outlining five recommendations to improve ECO implementation that it believed EPA had the authority to adopt (CAAAC 1995, 6–7):

- State or regional plans: allow states or regions to assume some of an employer's responsibility by implementing state or regional trip reduction programs
- Emissions equivalence: allow employers to substitute equivalent emissions reduction in lieu of submitting a plan to increase vehicle occupancy
- Good faith efforts: retain the ability of individual states to define and recognize good faith efforts as a measure of compliance with the program
- Credit for all trip reductions: allow full credit for the reduction of any trips, whether work-related or not, and for the participation of any group, including driving-age students
- Seasonal plans: allow states, regions, or employers to implement seasonal plans

On August 18, 1995, EPA incorporated all of the CAAAC's recommendations into a detailed new guidance (USEPA 1995a) and allowed states to implement more flexible ECO programs. However, problems immediately surfaced for the EPA. In 1993 Maryland and Illinois had asked EPA to allow them to implement seasonal programs to meet their ECO obligations and had been told that seasonal programs would not satisfy the law. Two years later, EPA was now telling states that these kinds of programs were acceptable. This reversal struck some state officials as evidence that EPA did not know what it was doing. EPA's exercise of administrative discretion also struck some members of Congress as contradicting the law's original intent. Manzullo questioned whether EPA had the legal authority under the CAAA-90 to permit regional programs and especially seasonal programs.

EPA's effort to give states more flexibility failed to stop the criticism and save ECO in its original formulation. Opposition to the program had already developed a critical mass. By 1995 many affected states had formally or informally suspended their ECO regulations. Employers had convinced themselves that the ECO program was costly and inconsequential and would not be enforced; they either stopped or scaled back their implementation.

Congressman Manzullo continued to promote his bill and by mid-December had signed on 159 cosponsors, including many seasoned Democratic legislators. Using the Consent Calendar reserved for noncontroversial measures, Manzullo's bill came up for a House vote on December 12, 1995. Waxman unexpectedly rose "to congratulate the gentleman from Illinois for this legislation. The administration, to my knowledge, has expressed no opposition to this legislation. I would urge the President to sign the bill" (U.S. Congress, House 1995b, 14269). With this support from an architect of the original bill and key member of the quality of life advocacy coalition, the bill passed the House. Two days later the Senate passed the bill by unanimous consent. President Bill Clinton signed the bill on December 23, 1995, and it became Public Law 104-70. Thus, ECO was reformulated from a mandatory to a voluntary program.

Although the economic development advocacy coalition may have hoped that they had heard the last of ECO, it was not to be. Manzullo's bill was "emissions neutral." It required states that ended their ECO programs to make up ECO's expected emission reductions from other sources. While urging his colleagues to support the bill, Waxman noted that "states can retain their ECO programs. Indeed, I fully expect that many of these programs will be retained. A well-designed and well-run ECO program can provide not only emissions reductions, it can reduce traffic congestion, give employees more commuting options, and encourage employer participation in regional transportation planning" (U.S. Congress, House 1995b, 14269). States took him up on the challenge. Even states that opposed mandatory ECO began to look more closely at voluntary ECO as a policy instrument for coming up with new sources of emission reductions. Some states already had voluntary programs in place, which appeared reasonable.

ECO policy continued. The perception and definition of urban air pollution caused by driving had not changed. Commute reduction as a policy instrument in the battle against air pollution remained on the systemic and institutional agendas. Despite opposition from several fronts, ECO lives on in legislative edicts and administrative guidelines as a reformulated policy. In the next section we examine the "new" commute reduction policy, voluntary ECO.

Reformulating Policy: A New Program Out of the Ashes

Removing Mandates: Making ECO Voluntary

People expect policy implementation to take months or even years, but they are often surprised to learn that a reformulated policy may take equally long to implement fully. Several factors impeded states from moving quickly to implement the reformulated ECO. First, states had their own due process requirements for removing regulations and amending statutes.[4] Second, EPA had to write new guidelines for states to remove ECO from state implementation plans (SIPs). Reformulating ECO did not remove the states' other clean air obligations. As Waxman stated earlier, "The bill is emissions neutral. It requires States that opt-out of the ECO program to make up the emission reduction from other sources" (U.S. Congress, House 1995b, 14269). States that had claimed emissions reduction credit from an ECO program in their SIP had to find some alternative way of making up the emission reductions. Even states that had taken no credit for emission reductions for ECO programs in their SIPs discovered that new EPA guidance for voluntary ECO required them to estimate what the reductions would have been under mandatory ECO. Then they had to show how they would make up that amount. This caused states to postpone the formal removal process until they had selected options for achieving an emission reduction equivalence.[5]

In fact, several groups claimed that voluntary ECO programs were or could be as effective in improving air quality as the more onerous mandatory efforts. Unfortunately, very few hard data exist on voluntary programs, and many people express skepticism about their abilities to clean the air. Research in other environmental areas, however, suggests that voluntary programs can accomplish more than mandatory ones among those individuals already committed to the goals of the program. However, voluntary programs accomplish less than mandatory ones among those individuals not committed to the program's goals (Scholtz 1991).

Four voluntary ozone control conferences were held between 1994 and 1997. These conferences supported the economic development coalition's instrument belief that voluntary programs should be granted SIP credits. A large number of people, representing both attainment and nonattainment areas, attended the conferences. EPA representatives spoke at all four conferences and found participant interest solidly in favor of granting credits for voluntary programs. Participants also wanted a system for forecasting and quantifying emission reductions from voluntary programs and the creation of a national clearinghouse for information on voluntary programs. The conferences provided an important forum for sounding out new ideas and establishing a solid basis of support for voluntary commute reduction. Serious efforts were also under way to rigorously and systemically evaluate voluntary programs. Several states, including California, had begun formal policy evaluations to determine how well voluntary programs measured up to their mandatory predecessors.[6] EPA was looking for just this kind of definitive information. Although the results of the first evaluations would not be available until 1998, the methodologies used could help other states determine whether voluntary programs produce measurable and positive results.

Episodic Programs: An Old Idea Gets a New Look

Although EPA initially refused to allow states to implement seasonal or episodic commute reduction programs as part of their required ECO plans, the reformulation of ECO piqued a new interest in these kinds of programs. High levels of ozone usually occur in the summer, when hot, stagnant air is trapped over an urban area for several days. States questioned why they should implement ECO programs during the winter when little ozone is produced or on those summer days when the weather naturally disperses the ozone. Further, the CAAAC had recommended in 1995 that affected states be allowed to implement seasonal plans as one of their options.[7]

In May 1996, EPA's Office of Mobile Sources commissioned a private contractor to evaluate mobile source episodic control programs.[8] The evaluation was divided into sections. The first section was completed in April 1997 and gave a descriptive snapshot of the more than thirty episodic and seasonal programs already in existence throughout the country. The description focused on each program's design, emission reductions, costs, public education efforts, resources, and evaluation results. The second section (completed in fall 1998) provided a detailed assessment of five programs in different parts of the country that show great variety in their episodic programs.

Figure 8.1 shows thirty-five episodic control programs, but only twenty-eight specifically target ozone (O_3) as their main pollutant; the rest target carbon monoxide (CO) or particulate matter (PM). Although programs in San Francisco and Tulsa date back to 1991, the majority began in 1994 or later. All of the ozone programs are voluntary and operate only during the summer season. All of the PM and CO programs are mandatory and operate only during the winter months.

These episodic programs are used in both attainment and nonattainment areas, but the motivations behind the programs differ in the two areas. Attainment areas hope the programs will help them maintain their status. Nonattainment areas hope to reduce or limit the number of air quality violations to help move toward attainment.

The policy impact expected from most voluntary programs is a reduction in the number of air quality violations. One of the policy outputs, how-

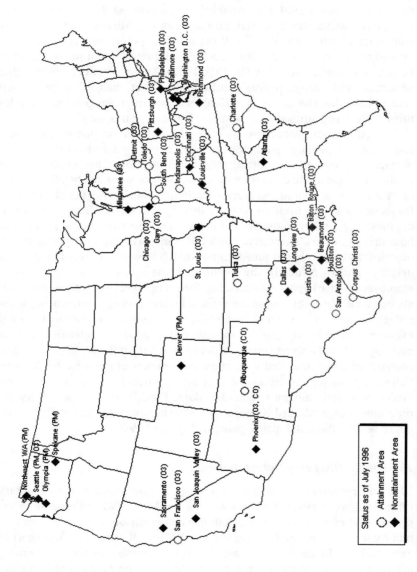

Figure 8.1 Episodic Control Programs in the United States

Status as of July 1996

○ Attainment Area

◆ Nonattainment Area

Northwest WA (PM)

Seattle (PM, O3)

Olympia (PM)

Spokane (PM)

Sacramento (O3)

San Francisco (O3)

San Joaquin Valley (O3)

Phoenix (O3, CO)

Albuquerque (CO)

Denver (PM)

Milwaukee (O3)

Chicago (O3)

Gary (O3)

St. Louis (O3)

Tulsa (O3)

Dallas (O3)

Longview (O3)

Austin (O3)

San Antonio (O3)

Corpus Christi (O3)

Houston (O3)

Beaumont (O3)

Baton Rouge (O3)

Detroit (O3)

Toledo (O3)

South Bend (O3)

Indianapolis (O3)

Cincinnati (O3)

Louisville (O3)

Pittsburgh (O3)

Philadelphia (O3)

Baltimore (O3)

Washington D.C. (O3)

Richmond (O3)

Charlotte (O3)

Atlanta (O3)

ever, is a more informed public. Unlike mandatory ECO, which devoted lit-tle time and few resources to educating the public about commute reduc-tion, the voluntary programs emphasize public education. In fact, a majori-ty (53 percent) of voluntary program coordinators listed public education as their number one goal. Only one-third (30 percent) said their most impor-tant goal was to attain or maintain air quality. Most of the remainder emphasized health benefits. A well-orchestrated public outreach effort was viewed by all areas as a prerequisite for public interest and participation in voluntary episodic programs (USEPA 1997). These programs assume that education will change people's attitudes and that changed attitudes will change behavior. Our findings in Chapter 6, however, suggest that the rela-tionship between attitudes and behavior is not quite this simple.

Since episodic programs are designed to increase public awareness and, ultimately, change public behavior on days of poor air quality, all the programs encourage widespread public participation. They encourage the public to take specific actions on ozone alert days, when the weather condi-tions are forecast to trap and convert emissions into ozone. People are most frequently asked to limit automobile driving and refrain from using small engines such as lawnmowers. People are also encouraged to avoid rush hour driving, to use alternative modes of transportation, to combine multi-ple trips, to keep vehicles tuned, to take lunch to work, to avoid lunch hour driving, and to refuel in the evening. Most areas (86 percent) asked employers to participate by educating their employees on actions individu-als can take, changing employee work schedules, using conferencing tech-nologies instead of face-to-face meetings, postponing fleet refueling until evening, and reducing high ozone-emitting production activities such as painting and landscaping activities. More than half of the areas (57 percent) encouraged companies that were stationary sources of pollution to also par-ticipate by postponing certain high-emitting production, maintenance, and landscaping activities on ozone alert days. Participation in these episodic programs is high; about 150 employers and stationary sources in Baltimore, out of several thousand, participated in 1995 and 1996.

Emission Credits: Policy Output or Impact

EPA was reluctant to consider granting emission credits for voluntary efforts for two reasons. The first was the need for valid and reliable meas-urements to effectively monitor both trip and emission reductions. The sec-ond reason was the need for backup plans if actual reductions fell short of expectations. In short, EPA wanted realistic measurements of voluntary efforts and firm assurances that a mandatory backup plan could be imple-mented quickly to make up any differences between expectations and results. Despite their popularity, voluntary episodic programs usually lack the measurable emission reductions required by the Clean Air Act. Voluntary episodic programs cannot ensure that the public will participate, nor can they ensure that a change in public behavior will be sufficient to register any improvement in air quality. Furthermore, if improvement in air quality is measured, it is difficult to prove that these programs caused the improvement.

Nevertheless, EPA, spurred by congressional actions, began to seriously explore the possibility of granting emission credits for voluntary programs. In 1996, Barton introduced an amendment to the CAAA-90 that included a section granting states credit for episodic programs. Even though this bill had limited prospects of making it through Congress, EPA thought it made sense to revisit the issue. Many voluntary episodic programs existed, and information about the programs was accumulating. Getting advocacy coalitions to promote and fund voluntary programs would be easier if credits were available at the end of the road.

On October 23, 1997, EPA made a major policy reversal and granted emission reduction credits for voluntary mobile source programs submitted in SIPs. The policy applied to episodic, seasonal, and continuous programs. Programs claiming credits were initially limited to five years, after which EPA would decide whether the expected reductions in air pollution had been met and whether the credits could continue. The new policy applied only to voluntary mobile source programs, and EPA made it clear that it was not establishing a precedent for credits that would apply to other emission sources. Finally, EPA set a limit of 3 percent on the emission reductions that an area could claim because "of the innovative nature of voluntary measures and EPA's inexperience with quantifying their emission reductions" (USEPA 1997, 9). EPA's own analysis of several voluntary programs revealed that the typical emission reduction was a fraction of 1 ton per day. Therefore, a limit on voluntary programs of 3 percent of an area's total reductions was reasonable.

EPA made the requirements for taking advantage of the new policy quite specific and substantial. These requirements included programs for which program outputs and emission reductions could be quantified and measured. States were obliged to monitor, assess, and report on program implementation and be prepared to remedy any emission reduction shortfalls quickly. Thus, if voluntary programs did not achieve their expected emission reductions, the state had to have a mandatory backup plan.

EPA provided air quality areas with an information clearinghouse. EPA's Web site includes detailed information on all of the existing voluntary programs and the methods they use to quantify emission reductions. A blueprint for program design outlines successful program design options for voluntary programs. It remains to be seen if this information will be enough for air quality regions to meet EPA's criteria for SIP approval. To date, no state has tried to get credit for its voluntary programs.

Is Commute Reduction in Our Future?

Cars still make a substantial contribution to urban air pollution, road congestion is increasing, and the automobile is dominating urban space. All three problems could be reduced by changing commuting behavior. However, the fact that opposition to ECO caused it to be reformulated before it could be fully implemented suggests that selling solo commute reduction to the public as a solution to these problems will be difficult. So, what lies in the future for ECO or similar policies?

Our research on ECO suggests the value of integrating policy subsystems to deal with the air pollution and traffic congestion problems in urban areas. These and other problems are increasingly interconnected, and efforts to resolve them must coordinate activities in several policy subsystems. Commute reduction policies need substantial involvement of the transportation and clean air subsystems, and the urban subsystem should not be left out of attempts to come up with solutions for problems facing urban areas. The transportation subsystem has a great deal of experience and expertise in the efficient movement of goods, services, and people. However, the effect of transportation on air quality and the effect of roads and parking on the urban environment have not been major issues in the transportation subsystem. Solutions used for other problems in the urban subsystem can be incorporated into clean air policies as well. New taxation polices can provide policy instruments that governments can use to discourage solo driving. Affordable housing is needed throughout the region, so people will not have to travel very far to jobs. Businesses need encouragement to locate near public transit routes, where they are accessible to most of their employees.

The role of advocacy coalitions is critical in understanding what happens to policy. ECO pitted the economic development advocacy coalition, represented by most businesses in the regions, against the quality of life advocacy coalition, represented by state agencies. Other members of the quality of life coalition took a low profile during the implementation stage. When voluntary commute reduction entered the picture, both advocacy coalitions embraced the idea, although for different reasons. The economic development coalition embraces voluntary commute reduction because it preserves individual freedom and does not hold businesses responsible for changes in their employees' nonwork behavior. The quality of life coalition embraces voluntary commute reduction because it still links changing people's behavior to improving the environment.

The effect of the intersection of policy subsystems can be clearly seen in discussions about congestion pricing. Unlike most toll roads that charge a single price, toll roads using congestion pricing charge higher rates at peak travel times than at low travel times. The rate can also vary with the type of vehicle and number of occupants. Although some actors in the transportation subsystem are promoting congestion pricing, actors in the other policy subsystems are also considering this option. The economic development coalition approves of any movement away from regulation and toward pricing mechanisms to solve air pollution and traffic congestion since market solutions are a favorite policy instrument in its belief system. The quality of life coalition may also accept the utility of congestion pricing, especially if the money made from charging people to use the asphalt commons is used to improve public transportation. However, poor people, already hurt most by automobility in the United States, will be further disadvantaged and restricted in their travels by congestion pricing unless public transit alternatives to the car are increased.

Congestion pricing is being tried in various parts of the country. A private road in California, built down the middle of a public road, opened in December 1995. The privately funded road charges a toll for single occupancy vehicles based on congestion, but the public road remains free. Evidence shows reduced congestion on the public road and more people carpooling so they can use the new toll road. Other pilot projects have tried congestion pricing on public highways and bridges, but citizens have objected to paying to use bridges and highways already built and paid for with taxpayers' money (Marzotto 1998).

Policymakers are beginning to think in regional and global terms rather than just in terms of political units such as cities, states, or even nations. However, a major obstacle to regionalism is that public policy must still be formulated and implemented by governmental units, and few governmental units exists for regions. Most urban areas in the United States are contained within the nation's boundaries, but many sprawl across state borders. Attempts to govern these multistate cities with direct federal or regional policies run squarely into state sovereignty, something jealously guarded by state officials. Furthermore, states assign responsibility for many governmental functions to cities, counties, and other local jurisdictions, who also carefully guard their turf. The boundaries of these jurisdictions, based on history rather than current conditions, are difficult to change. Regional solutions to problems created by automobility are hindered by the interests of state and local governments. These interests may need to be overridden if regional problems are to be solved with regional solutions. ECO is a prime example of a policy conceived of in regional terms, with the nonattainment areas in the CAAA-90 defined regionally but given to nonregional entities, states, to implement. States found it difficult to coordinate commute reduction efforts in multistate standard metropolitan statistical areas (SMSAs), such as the Philadelphia urban area. However, both the economic development and quality of life advocacy coalitions have begun promoting regionalism for other reasons. This may signal an external system change that would make regional policy instruments more acceptable to policymakers.

Air pollution and congestion are two problems in the urban commons that may be getting worse. Government regulation and market pricing are two very different ways to restrict overuse of the commons. Regulations rely upon force and changing people's behavior from the outside. Market pricing raises the cost of scarce commodities until the demand declines to meet the supply and focuses on changing people's behavior from the inside through personal choices on how to allocate their own limited resources. The price put on the use of the transportation commons should include the costs associated with its overuse—air pollution, traffic congestion, diminished quality of life—and the cost of restoring it. Market pricing could eventually change the use of the commons, but regulations would be needed to prevent irreversible destruction of the commons while pricing mechanisms take effect. Government regulations set the stage and begin the movement toward change, and market pricing continues that movement by encouraging private decisions that will contribute to the common good. The problem of ozone pollution and the contribution the car makes to it have not gone away. ECO may have failed as it was initially formulated, but that initial formulation may have set the stage to use the market to raise the cost of driving alone and thereby change driving behavior.

Conclusion

ECO did not end; it changed. In fact only a single word in the original CAAA-90 was changed, from "must" to "may," which moved ECO from a mandatory to a voluntary policy. However, the problems ECO was formulated to solve have not gone away. Areas with high levels of air pollution still face the challenge of improving their air quality. Areas with relatively clean air must still struggle to keep the air clean. The problems remain; only the solution has changed.

When Congress reformulated ECO, one of the first changes was that the EPA no longer required that states have year-round commute reduction programs. Instead, states were granted SIP credits for seasonal and episodic programs that encourage businesses and residents to cut back on pollution producing activities on bad air days. Attainment areas have a strong desire to implement voluntary episodic ECO programs since these programs may help them maintain their status. Reclassification from attainment to nonattainment brings stringent EPA regulations with significantly increased costs and mandated programs (e.g., vehicle inspection and maintenance procedures). Nonattainment areas are already subject to these EPA regulations; however, the gains, though small, from episodic ECO programs could be just enough to keep ozone levels in attainment areas at acceptable levels. The promise of even small gains in air pollution reduction, coupled with the prospects and advantages of changing driving attitudes and behavior, seems to have given many nonattainment areas the necessary encouragement to implement voluntary episodic ECO programs.

When a policy goes back through the cycle and is reformulated, a wider range of options is often considered than during the initial cycle. The original ECO policy focused so much on employees' driving behavior that it ignored larger areawide policy instruments, such as congestion pricing and regional car pool promotion. Whether any of these new programs will reduce the basic problem of pollution caused by too much solo driving is not yet known. Also unknown is whether the original mandatory ECO policy would have reduced the problem if fully implemented. However, we feel confident that a policy is more likely to be reformulated than terminated once it has been established as a solution, even an unsuccessful one, and that ECO will be around in one form or another for a long time.

The change in ECO was influenced by changes in the political climate, especially the election of a new systemic governing coalition. Had the Democrats retained control of Congress, ECO might have become one of those largely ignored public policies (like the 55 mph speed limit) or one with uncertain impact yet widely complained about. ECO was an easy target for the new Republican Congress to demonstrate its resolve to "rescue" states and businesses from "big government" and "unfunded mandates," and yet affirm its commitment to the environment. The change in ECO did not threaten the overall CAAA-90, maintaining the goal of clean air while eliminating a much disliked and burdensome policy for states and businesses.

Both advocacy coalitions were pleased with voluntary ECO, which encourages behavioral change as a means to a clean environment, while at the same time preserves individual freedom and personal choice. The quality of life coalition did not have to compromise its basic belief that people need to change their behavior to improve the environment. Reformulated ECO only required the coalition to give up its instrumental belief that regulations are required to change this behavior. Although the economic development coalition did not strongly object to the belief that people had to change behavior, it did object to a mandate being placed on businesses. It could accept a voluntary ECO policy as supportive of its basic beliefs about individual freedom and personal choice, its policy belief that government regulations should be as "business friendly" as possible, and its instrument belief that the government shouldn't impose unnecessary costs on businesses. Thus, the quality of life coalition accepts voluntary ECO because it retains the original definition of the problem, keeps the issue on the institutional agenda, and recognizes the need for individuals to change their driving behavior. The economic development coalition accepts voluntary ECO because it protects individual freedom, allows for personal choice in commuting decisions, and eliminates the mandate on businesses.

Once the policy cycle gets under way, policy activity becomes continuous, with fluctuations in the scope, intensity, and nature of the activity. The beginning of the policymaking process is marked by the emergence of a problem as individuals, groups, and advocacy coalitions define the problem and promote certain solutions based on their belief systems. Each wants their definitions and solutions placed on governmental agendas. Policy brokers often assist in formulating and adopting a policy. Several public and private entities implement the policy before the policy reaches the final target group. In the process, informal and formal evaluations of the policy are made, some claiming the policy has defects. Since polices are more likely to be altered than terminated, the policy most often returns to the legislature to correct these defects, and the cycle repeats itself frequently. Advocacy coalitions continue to influence each stage of the repeated cycles.

Policies evolve slowly. They don't start and stop abruptly. They often become settled for a time, handled by routine administrative procedures. Other times, somewhere along the way in the cycle, something changes. A problem may improve or worsen. The influence of advocacy coalitions may increase or decrease. Changes in socioeconomic conditions, public opinion, public behavior, or governing coalitions may raise new problems or redefine old ones. Even the implementation of an old policy may raise the call for change. All of these can send policies back through the policy cycle advocacy system. The policy system is cyclical, rather than linear, frequently repeating as policies are pushed through the various stages by advocacy coalitions working within and across policy subsystems that are part of a larger political, social, and natural system. New problems connect to old problems, their solutions become linked, and new policies are born.

Notes

1. Sen. Chafee did concede that "over a very long period, a requirement like this would convince major employers to make locational decisions that encourage

the use of transit and other ridesharing options. But in the short-run, the emissions reductions achieved do not justify the great difficulties that would be experienced by the States and by employers to carry out the trip reduction program" (U.S. Congress, Senate 1995, 18575).

2. In 1992 a study issued jointly by the City of Chicago and Roosevelt University reported that federal environmental laws would cost state and local governments $32 billion in 1995 alone. See Roosevelt University (1992).

3. Manzullo was supported in his efforts by the Chicagoland Chamber of Commerce, the Illinois legislature, and the governor. In January 1994 the Illinois House passed a joint resolution calling upon the governor and members of the Illinois congressional delegation to work with other states and their congressional delegations to amend the CAAA-90 (Ibata 1994e). The Illinois resolution was a significant first step in the state's opposition to ECO.

4. Some states such as New Jersey had codified their ECO regulations into state law, which required legislative action to remove.

5. EPA's guidance suggested that some ECO substitutes might include

- measures not federally mandated,
- surpassing the requirements of federally mandated measures (i.e., going beyond required reductions or level of implementation), or
- newly identified or discretionary measures states choose to include in current SIP submittals.

6. In September 1996 the California state legislature passed S.B. 836, which was based on the premise that voluntary ride sharing may be as effective as mandatory programs. The new law required that voluntary programs be formally evaluated to see if their levels of emission reductions were comparable to those anticipated for mandatory programs. The legislation also required the South Coast Air Quality Management District to provide $1.5 million annually to the Regional Transportation Agencies Coalition to promote ride sharing during the morning commute rush hour. The first evaluation results were expected sometime in 1998 (Hawthorn 1997, 10–11).

7. Seasonal ozone, carbon monoxide, and particulate programs operate during a region's high pollution period. For example, regions in the Pacific Northwest operate carbon monoxide programs during the winter season but ban burning throughout the year. Episodic programs operate only on days when there is evidence of high levels of pollution. Episodic programs may also be seasonal. In the Northeast, many ozone reduction programs operate during the summer (seasonal) on days with high ozone alerts (episodic).

8. EPA's decision to study emission reductions from these episodic programs was partly preempted by the introduction in 1996 of a bill (H.R. 3519) in the House of Representatives to amend the Clean Air Act. The bill was introduced by Joe Barton (R-Texas), chair of the House Energy and Commerce Committee's Subcommittee on Oversight and Investigation, and included a provision for crediting voluntary programs. Barton chaired the hearings on the Manzullo bill in 1995. The bill included a provision requiring the administration of EPA "when evaluating the adequacy of State implementation plans for national primary and secondary ambient air quality standards, to credit provisions designed to control air pollution only during certain periods during which pollutant levels are elevated" (U.S. Congress, House 1996, 5569).

Appendix

Study Participation

Companies

Our study included a sample of companies in the Philadelphia air quality region that would be affected by employee commute options (ECO) policy. Our basic sample came from a database consisting of private companies with 100 or more employees in the City of Philadelphia, the five surrounding Pennsylvania counties, six New Jersey counties, and the two Delaware counties that make up most of the Philadelphia air quality region. (Cecil County, Maryland, was omitted because of the small number of large businesses in the county.) This database was obtained from Dun and Bradstreet, a private company that compiles company lists. We randomly sampled from within thirty-six strata formed by classifying companies by four geographical areas (City of Philadelphia, Pennsylvania counties, Delaware counties, and New Jersey counties), three size groups (100–499, 500–999, and 1,000 or more employees), and three industry groups (manufacturing, service, and other).

We supplemented this basic sample with two additional samples. The first supplement included a set of Delaware companies that were not included in the basic sample. Three-fifths of the Delaware employers in the basic sample had either only a few employees in Delaware or no Delaware telephone listing. Additionally, half (thirty-two) of the private-sector members of the New Castle Transportation Management Association were not in the Dun and Bradstreet database. The supplemental Delaware sample included these thirty-two companies, seven of which agreed to participate in our evaluation.

A second supplement included Maryland companies primarily in the Baltimore air quality region, since the only Maryland county (Cecil) included in the Philadelphia region had few companies with 100 or more employees. These Maryland companies had said that they intended to set up voluntary ECO programs in anticipation of the delayed state regulations.

We contacted 479 companies in an attempt to recruit three companies in each of the thirty-six strata of the basic sample. Half of those we contacted did not have a site in the region with 100 or more employees, no longer had a telephone listing, or were otherwise not appropriate for our study. Of those eligible, ninety-one agreed to participate in our study. We recruited five from the supplemental list of thirty Delaware companies and fourteen from the supplemental list of thirty-two Maryland companies. A total of 112 companies (ninety-one from the basic samples and twenty-one from the supplemental samples) agreed to participate in the research, or 41 percent of the companies we contacted that had to implement ECO programs. (See Table A.1.)

Table A.1   Number of Private Companies by Study Involvement and Area

Study Involvement	DE	NJ	PA Sub.	PA Phil.	MD	Total
Dun and Bradstreet Basic Sample						
Number of employers listed	310	643	1,214	673	n.a.	2,840
Number with contact attempt	172	111	105	91	n.a.	479
No telephone listing	31	14	21	15	n.a.	81
Under 100 employees	68	30	37	22	n.a.	157
Public, duplicate, other	6	4	4	1	n.a.	15
Number eligible	67	63	43	53	n.a.	226
(percentage contacted eligible)	(39%)	(57%)	(41%)	(58%)		(47%)
Agreed to participate	16	29	28	18	n.a.	91
(response rate)	(24%)	(46%)	(65%)	(34%)		(40%)
Baseline employee	7	13	0	1	n.a.	21
Organization survey	6	11	7	8	n.a.	32
ECO plan	7	10	1	0	n.a.	18
Supplemental Samples						
Number of employers	30	1	n.a.	1	32	64
No phone listing	1	0	n.a.	0	1	2
Under 100 employees	1	0	n.a.	0	2	3
Duplicate, stop work, other	10	0	n.a.	0	1	11
Eligible with contact	18	1	n.a.	1	28	48
(percentage contacted eligible)	(60%)	(100%)		(100%)	(88%)	(75%)
Agreed to participate	5	1	n.a.	1	14	21
(response rate)	(28%)	(100%)		(100%)	(50%)	(44%)
Baseline employee	1	0	n.a.	0	8	9
Organization survey	3	1	n.a.	—	5	9
ECO plan	2	1	n.a.	0	0	3

The organization surveys, containing seven booklets of four different types, were sent to company ECO transportation coordinators in July 1995. The organization culture surveys were to be completed by three people in different levels within the companies: top management, midlevel supervisor (either line or staff), and a nonsupervisory employee (either staff or line but different from the midlevel supervisor). The transportation coordinator survey focused on costs and benefits, if any, of planning the ECO programs. The line executive survey focused on the goals of the company and the business climate in which it operated. The human resources executive survey focused on what was important within the organization and how the company started change. Surveys were returned by forty-four companies; some returned all seven of the booklets and others returned only one. (The forms are not reproduced since they provided little information for this book.)

A telephone follow-up survey was conducted in March 1996. It collected information from ninety-one transportation coordinators about the status of their ECO programs, and why they had or had not planned and started a program. (See Figure A.1.) Characteristics of the companies were verified also.

Employees

Employee surveys were completed by 4,787 employees from fifteen firms in the two air quality regions (Philadelphia and Baltimore). Companies distributed the questionnaire to employees who were scheduled to report to work between 6:00 A.M. and 10:00 A.M. during a specified five-day work week. Employees answered questions regarding commuting habits, attitudes toward protecting the environment, current behavior related to the environment, the team atmosphere of the work setting, and socioeconomic characteristics. (See Figure A.2 for the Employee Survey Form.)

Analytic Measures

Companies

Cost measures came from the organization survey's transportation coordinators form. Transportation coordinators were asked to provide as much detail as possible about the costs of their ECO programs. Costs associated with obtaining information, developing program infrastructures, and planning were combined into planning costs. Costs associated with promotion, management, incentives, capital investments, and schedule changes were combined into implementing costs.

The compliance measures came from the telephone follow-up survey. Planning compliance is the number of ECO components that companies included in their plans. Half the companies reported no planned components, whereas two reported fifteen. Treating planning compliance as a dichotomous variable made little difference in the regression shown in Table A.2. It did eliminate the effect of the human resources director being the ECO transportation coordinator. Overall compliance ranged from 0 to 3, with companies receiving a "0" if they did not develop an ECO plan, a "1" if they made a plan but did not implement any of it, a "2" if they implemented part of their plan but stopped when the state no longer required implementation, and a "3" if they voluntarily planned to continue with at least part of their ECO plan.

**Table A.2   Company Compliance with ECO
(standardized multiple regression coefficients)**

Company Characteristics	Planning Compliance	Overall Compliance
New Jersey	0.59*	0.40*
Delaware	0.38*	0.43*
TMA membership	0.25*	0.23*
Employs 750 or more	0.20*	0.23*
Employs 350–749	0.04	−0.13
ECO coordinator as human resource manager	−0.19*	−0.01
Importance of overall image	−0.15	−0.11
Importance of green image	−0.16	0.04
Near transit stop	0.07	0.02
Services company	0.12	−0.09
Manufacturing company	0.07	−0.05
Privately held for-profit company	−0.09	0.03
Nonprofit organization	−0.04	0.20
Branch site of company	−0.04	0.08
Union representation	0.05	0.02
Multiple daily shifts	−0.06	−0.12
R^2	0.55	0.50

*Statistically significant ($p < 0.05$).

Overall image is a scale that came from a question (Q4) human resource executives were asked: "How important are the following goals to your organization's top management in making strategic decisions, or commitments of a long-term nature?" Nine items were listed under this question, with response categories ranging from (1) "moderately important" to (7) "extremely important." Two of the items clustered together in a factor analysis, and the resulting scale had high reliability (Chronbach's alpha = 0.92):

- achieving or maintaining a high (above average) public image
- retaining or securing a high (above average) reputation as a good community citizen

Green image is a scale that came from line executive responses to the same question as the human resource executives answered, except line executives were given only three items of the nine items to rate, which clustered well into a single scale (Chronbach's alpha = 0.91):

- producing "green" products that will protect the environment
- achieving or maintaining a reputation for environmental sensitivity or a "green" image
- committing sufficient resources to the production process to meet or exceed environmental protection standards

Employees

The average daily car contribution is the number of vehicles brought to work by the employee during the survey week, divided by the number of days worked by the employee during the week. (See Table A.3.) A score of 1.0 means an employee drove alone to work each day worked. A score of 0.5 shows the employee was in a two-person carpool each day and assumed to drive once every two days. A score of 0.0 means that the employee walked, bicycled, or took public transit to work each day. Other intermediate scores show the employee came to work one way on some days and another way on other days. An employee who telecommuted was counted as working but bringing no vehicles to work on that day. An employee who worked a compressed work week was counted as working on the compressed day off but bringing no vehicles to work. The average daily car contribution, which can be measured for an individual, is the inverse of the average vehicle occupancy (AVO) and average passenger occupancy (APO), which are group measures specified in ECO requirements. The data came from employee answers to Q2 through Q13. (See Figure A.2.)

The Carpool orientation scale was developed from three questions about carpooling (Q23, Q24, and Q25). These clustered in a single factor with good scaling reliability (Chronbach's alpha = 0.74). The direction of the coding was reversed from that of the individual questions so that the higher the scale value, the more favorable the attitude toward carpooling.

Two environmental attitude scales derived from the agreement with fifteen statements about humans and the environment (Q27). Twelve of these

Table A.3   Regression of Values, Attitudes and Behavior on Each Other and Background Characteristics (standardized regression coefficients)

Independent Variable	Environmental Sensitivity	Man Over Nature	Carpool Orientation	Environment Actions	Average Daily Car Contribution
Carpool orientation	—	—	—	—	−0.23**
Environment actions	—	—	0.06*	—	0.08**
Environment sensitivity	—	—	0.19**	0.27**	−0.04
Man over nature	—	—	−0.02	−0.05*	−0.03
Current teamwork	−0.00	0.07**	0.09**	0.06*	0.01
Office relationships	0.04	0.04	0.02	0.04	0
Outside relationships	−0.01	0.07**	0.07**	0.06*	−0.02
Ideal teamwork	0.18**	−0.08**	0.01	0.03	0
Cars per HH member	−0.02	−0.02	−0.07**	0.06**	0.26**
Education	0.02	−0.03	−0.05*	0.03	0.06**
Age	0.07**	0.02	−0.04	0.11**	0.02
Family income	−0.06*	−0.04	−0.09**	0.01	0.01
Commute distance	0.03	0.01	0.02	0.02	−0.01
Male	0.01	0.07**	0.01	0.08**	0.01
R^2	0.04	0.02	0.11	0.13	0.16

* $p < 0.01$ ** $p < 0.001$

came from the Riley Dunlap and Kenneth Van Liere's new environmental paradigm (NEP) scale (1978); the other three were developed for this study. Factor analysis identified two factors. One item (Q27a) did not load very strongly on either factor and was dropped from further analysis. Environmental sensitivity had high-scale reliability (Chronbach's alpha = 0.85). Man over Nature had lower but still acceptable scale reliability (Chronbach's alpha = 0.66).

Environmental actions was derived from seven environmental behaviors that employees could do "never" to "all the time" (Q28). We added to Daniel Sivek and Harold Hungerford's (1990, 36) definition that environmentally responsible behavior is "any action, taken by an individual or group, directed toward the remediation of environmental issues or problems." This scale included simply "doing" something out in nature. All seven items clustered into a single scale with good scale reliability (Chronbach's alpha = 0.78).

The teamwork and relationship scales came from twenty questions (Q31 and Q32) derived from the work of Dean Tjosvold (1984) and Johnson et al. (1981). Factor analysis found these questions clustered into four scales and converged in a verimax rotation in seven iterations. Ideal teamwork included seven items with a Chronbach's alpha of 0.91 (Q32a, Q32c, Q32d, Q32f, Q32h, Q32i, Q32j). Current teamwork included the equivalent seven items from Q31 with a Chronbach's alpha of 0.90. Office relationships included four items (Q31e, Q31g, Q32e, Q32g) with a Chronbach's alpha of 0.79. Outside relationships included two items (Q31b and Q32b) with a Chronbach's alpha of 0.65.

The likely influence of company programs to reduce rush hour commuting comes from responses to Q29 and Q30. Employees rated each of the fourteen items in Q29 from not very influential to very influential. They rated each of the six items in Q30 from not very justified to very justified. Factor analysis identified four factors that converged in six iterations of a verimax rotation. The schedule programs scale included three items (Q29i, Q29j, Q29k) with a Chronbach's alpha of 0.81. The carpool programs scale included four items (Q29a, Q29b, Q29c, Q29d) with a Chronbach's alpha of 0.81. Transit programs included seven items (Q29e, Q29f, Q29g, Q29h, Q29l, Q29m, Q29n) with very high scaling reliability (Chronbach's alpha = 0.92). The negative sanctions scale involved the six items of Q30 with a Chronbach's alpha of 0.91.

Cars per household member was produced by dividing the number of vehicles in the home (Q22) by the number of people in the home (Q21). The survey did not include a question on the number of licensed drivers in the home, which would have given a better measure of the availability of vehicles for commuting to work.

Figure A.1   Economic and Social Effect of Employee Commute Regulations

Employer Follow-Up Survey

Center for Suburban and Regional Studies
Towson State University
February 27, 1996   STUDY ID: __/ __/ - __/ __/ - __/

Company Name: _____
Contact Name: _____
Telephone Number: _____

STATUS FILE INDICATES BASELINE SURVEY:	YES	....	1
	NO	....	2
STATUS FILE INDICATES STATE PLAN:	YES	....	1
	NO	....	2

CALL ATTEMPTS			
DATE	TIME	RESULTS	INITIAL

Comments: _____

FINAL STATUS:		
	Completed interview............................	1
	Refused interview	2
	No contact with appropriate person after 6 attempts	3
	No appropriate person identified for interview.........	4
	No contact with company after 6 attempts	5

REVIEW *COVER SHEET* AND *STATUS FILE PRINTOUT* BEFORE THE *INTER-VIEW*, CHECKING THE APPROPRIATE BOXES. IF *COVER SHEET* HAS INFORMATION NOT REFLECTED ON THE *STATUS FILE PRINTOUT*, MARK THE CHANGE ON THE *STATUS FILE PRINTOUT*.

My name is _____ of the Center for Suburban and Regional Studies at Towson State University. With all the discussions and changes in the Employee Commute Options or Employee Trip Reduction program, we have a few questions to ask you that will help us in the next stage of our research.

1. STATUS FILE INDICATES BASELINE SURVEY: We have an indicator in our file that your company has conducted a baseline employee survey (before 1996). Is this correct?

> Yes 1
> No... 2

STATUS FILE DOES NOT INDICATE BASELINE SURVEY: Did your company or organization conduct a baseline employee survey in 1993-1995 as part of ECO or ETR?

> Yes 3
> No... 4

2. STATUS FILE INDICATES STATE PLAN: We have an indicator in our file that your company filed an ECO or ETR plan with the state in 1994 or 1995? Is this correct?

> Yes 1
> No... 2

STATUS FILE DOES NOT INDICATE STATE PLAN: Did your company or organization file an ECO or ETR plan with your state in 1994 or 1995?

> Yes 3
> No... 4

3. Did your company or organization implement any part of an ECO/ETR program in 1995?

> Yes 1
> No... 2

4. Does your (company/organization) plan to continue or implement any part of an ECO/ETR program in 1996?

> Yes (CONTINUE) 1
> No (SKIP TO 5) 2

4A.   What were the primary reasons your company plans to start or continue ECO/ETR type programs?

    Positive Reasons

It benefits the company financially	1
It benefits the green image of company	2
Being a good neighbor in community	3
It's the right thing to do	4
Employees expect or want it	5
Assist employees in their work commute.	6
Fosters better employer-employee relationships	7

    Negative Reasons

Could be fined if we did not comply	11
State law or regulations require it	12
Already began, might as well continue	13
Might prevent worse policies.....................	14

4B.   Do you think your company or organization will continue to implement an ECO/ETR type program if the state makes it voluntary?

Yes (SKIP TO 6)................................	1
No (SKIP TO 6)	2

5. What were the primary reasons your company (did/does) not plan to start or continue any ECO/ETR program?

Too much trouble	1
Too intrusive in employees' privacy.	2
Too expensive.................................	3
Not worth the cost	4
Can't change employees commuting habits...........	5
Expected benefits for the environment are negligible ...	6

ASK 6 IF THE COMPANY FILED A PLAN (Q.2 CODED 1 OR 3) OR IMPLEMENTED A PROGRAM IN 1995. OTHERWISE GO TO 7.

6. We would like to know if what you (planned/implemented/plan to implement) included different programs. I could send you the list of the 21 programs and then call you back, or I could read each program to you and you could tell me if you (included a program of this type in your plan/implemented it in 1995/plan to implement it in 1996). Which would you prefer?

Read now (CONTINUE)	1
Send and call back (SKIP TO 7)...................	2

6A.   As I read each one, you can just tell me "Plan," "1995," and/or "1996."

		PLAN	1995	1996
1)	A transportation center or booth...........	I_I	I_I	I_I
2)	Transportation fairs, orientations, or meetings	I_I	I_I	I_I
3)	Personalized trip planning	I_I	I_I	I_I
4)	Ride matching or carpool locator services...	I_I	I_I	I_I
5)	Preferential parking for carpools	I_I	I_I	I_I
6)	Parking charges for solo commuters	I_I	I_I	I_I
7)	Vanpooling program....................	I_I	I_I	I_I
8)	Guaranteed ride home	I_I	I_I	I_I
9)	Bus or train schedules, maps, and other information	I_I	I_I	I_I
10)	Sell transit tickets or passes	I_I	I_I	I_I
11)	Subsidize or discount transit costs.........	I_I	I_I	I_I
12)	Shuttle service to transit stops, other work sites, or to nearby commercial and office centers..	I_I	I_I	I_I
13)	Flexible work hours	I_I	I_I	I_I
14)	Moving start times before or after rush hour .	I_I	I_I	I_I
15)	Compressed work week (fewer but longer days)	I_I	I_I	I_I
16)	Allowing employees to work at home some days	I_I	I_I	I_I
17)	Fitness programs for walking, running, or bicycling	I_I	I_I	I_I
18)	Showers, lockers, racks, or other facilities for walkers, runners, or bikers..............	I_I	I_I	I_I
19)	Commercial services on-site	I_I	I_I	I_I
20)	On-site or nearby daycare facilities	I_I	I_I	I_I
21)	Prizes, rewards, lotteries, or raffles	I_I	I_I	I_I

6B.   Did you plan or implement other types of ECO/ETR programs, and if so, what were they?

No or none other............................... 0

1

---

7. At this time, does your company or organization plan to conduct an ECO/ETR employee survey in 1996?

Yes (CONTINUE) 1
No (SKIP TO 8) 2

7A.   Would you be interested in Towson State University helping you conduct that survey?

Yes (CONTINUE) 1
Maybe, depends (CONTINUE) 2
No (SKIP TO 8) 3

Comments:

_____

_____

_____

7B.   In what month are you likely to conduct your next employee survey?

January or February 2
March 3
April 4
May... 5
June.. 6
July 7
August 8
September..................................... 9
October....................................... 10
November..................................... 11
December 12

8.   At this time, does your company plan to file an ECO/ETR plan or report with the state in 1996?

Yes 1
No.. 2

9.   The Center for Suburban and Regional Studies canceled a January workshop on the ECO/ETR program and our research findings because of limited registration. However, some of our participating companies were very disappointed and have asked if we could plan another one. How likely would it be that you or another person at your company would attend a half-day workshop on our research findings? Would you say...

Very likely (CONTINUE) 1
Likely (CONTINUE) 2
Unlikely (SKIP TO 10)......................... 3
Very unlikely (SKIP TO 10)..................... 4

9A.   We will let you know as soon as we have made plans. Since we are located in Baltimore but would want to have this workshop convenient to our companies in the Philadelphia region, would you have a suggestion as to a location?

_____

_____

10. |__| COMPANY IMPLEMENTING ECO/ETR (YES TO Q.4): Our research team wants to understand the specific experiences of companies implementing ECO/ETR programs, which cannot be obtained through a short, standardized questionnaire. Would it be OK for us to call in about a month to arrange a meeting between several of our team members and yourself?

Yes: Thank you very much for your cooperation. We will call you
      in about a month. Meanwhile, feel free to call us if we can
      help you in any way.   1

No: Thank you very much for your cooperation in our research.
      This concludes our data collection, but feel free to call us
      if there is a way we can help you.   2

|__| COMPANY <u>NOT</u> IMPLEMENTING ECO/ETR BUT MAY CONDUCT
      EMPLOYEE SURVEY WITH OUR HELP (NO TO Q.4 AND YES OR
      MAYBE TO Q.7A): Thank you very much for your cooperation. We will
      be back in touch about your employee survey in (MONTH BEFORE
      SURVEY MONTH). Meanwhile, feel free to call us if we can help you
      in any way.   3

|__| COMPANY <u>NOT</u> IMPLEMENTING ECO/ETR NOR EMPLOYEE
      SURVEY (NO TO Q.4 AND Q.7A): Thank you very much for your
      cooperation in our research. This concludes our data collection, but feel
      free to call us if there is a way we can help you   4

EMPLOYEE COMMUTE OPTIONS SURVEY
FOR THE FEDERAL CLEAN AIR ACT AMENDMENTS OF 1990

Developed and Conducted by:

CENTER FOR SUBURBAN & REGIONAL STUDIES
TOWSON STATE UNIVERSITY
TOWSON, MARYLAND 21204-7097
(410) 830-3827

COMPANY NAME

WORK SITE

MARKING DIRECTIONS

- USE NO. 2 PENCIL ONLY
- Fill the rectangle completely.
- Do not make any stray marks on this form.
- Erase completely to change a response.

CORRECT MARK

INCORRECT MARKS

IMPORTANT — **To the person filling out this form.** An Optical Mark Scanner will be used to collect the data off of this form for processing. We kindly ask that you DO NOT bend, fold or mutilate this form in any way. Thank you.

EMPLOYEE I.D. NUMBER   WORKSITE #

1. Date completed survey: ___ ___ / ___ ___ / ___ ___

These questions refer to the 5 week days ending last Friday (NJ–7 days ending last Sunday). Mark a code in the number for each question in the column unless the instructions say to *skip*. Each column is for a different day.

2. **Work status:**
   a. Worked at regular site
   b. Worked at another site
   c. Not scheduled to work *(SKIP Q.3-8)*
   d. Did not work, other *(Skip Q.3-8)*

	LAST WEEK						
	MON	TUE	WED	THU	FRI	SAT	SUN
						NEW JERSEY ONLY	

3. **Scheduled starting time:**
   What is your scheduled starting time?

   EXAMPLE
   HR   MIN
   6 : 30

   Fill in time and mark bubble for a.m. or p.m. for each day.

4. **Stopping Time:**
   What is your scheduled stopping time?

   Fill in the time and mark bubble for a.m. or p.m. for each day.

SCANTRON FORM NO. F-7168-TSU   © SCANTRON CORPORATION 1994 ALL RIGHTS RESERVED.   P4 3494-C E3515- 5 4 3 2 1

MAKE NO MARKS IN THIS AREA

0014001

CONTINUE TO NEXT PAGE

	LAST WEEK						
	MON	TUE	WED	THU	FRI	SAT	SUN
						NEW JERSEY ONLY	

5. Number of vehicles used in getting to work

6. Way of getting to work: Mark all the ways in column A, the way that covered the most miles in column B, and any way that did not come within 2 miles of your worksite in column C.

Private vehicle
Employer's vehicle
Vanpool (7 or more seats)
Motorcycle
Taxi
Bus *(Skip Q. 7-8)*
Train *(Skip Q. 7-8)*
Walk, run, bicycle *(Skip Q. 7-8)*
Worked at home *(Skip Q. 7-8)*
Other *(Specify_____; Skip Q. 7-8)*

7. Number of persons in vehicle, including driver *(Mark one bubble for each day.)*

a. Number going to your worksite

b. Number going to a different worksite

c. Number of children going to daycare facility

d. Number going full-time to post-secondary school

e. Number going to an adult daycare facility

f. Number going elsewhere or returning home

8. If you used a non-polluting or alternative fuel vehicle to travel to work, record the type:
a. Electric
b. Propane
c. Compressed natural gas
d. Methanol

9. Do you have a disability that requires you to drive alone in your own vehicle?
☐ No
☐ Yes *(Specify_____)*

10. What type of employment relationship do you have with the employer at this site?
☐ Permanent employee
☐ Temporary employee
☐ Employee of another company with contracts for work at this site

11. Do you have special arrangements to work longer days so you do not have to commute every day between 6 and 10 a.m.?
☐ No
☐ Yes, 36 hours in 3 days
☐ Yes, 40 hours in 4 days
☐ Yes, 80 hours in 9 days
☐ Yes, other *(Specify_____)*

12. Do you have special arrangements to telecommute so you do not have to commute every day?
☐ No
☐ Yes, telecommute from home
☐ Yes, telecommute from other location *(Miles from home to telecommute location _____)*

13. Do you have any other special arrangements to start work before 6 or after 10 a.m. so you do not have to commute during morning rush?
☐ No
☐ Yes

14. Home zip code:
a. Municipality

(New Jersey Only)

15. Number of road miles between home and work:

a. Number of these you drive alone:

16. How many minutes does it usually take to get from home to work?

CONTINUE TO NEXT PAGE

17. How much do you pay for parking?
(Fill in the amount and mark time period. Write "0" if none.)

$ [ ] [ ] [ ] [ ]

- ▭ Per day
- ▭ Per week
- ▭ Per month
- ▭ Per year

18. Do you have an assigned parking space? ▭ No ▭ Yes

19. While at work, about how often are you required to use a car for official purposes?
- ▭ Never *(Skip Q. 20)*
- ▭ Less than 1 day per week
- ▭ 1 day a week
- ▭ 2 days a week
- ▭ 3+ days a week

20. If you are required to use a car for official purposes, do you use a company car?
- ▭ No, none are available
- ▭ No, but I could
- ▭ Sometimes
- ▭ Most or all of the time

21. How many people live in your home? (Mark one.) ➤

CONTINUE TO #22

22. How many total vehicles do the people living in your home have? *(Mark one.)* ➤

23. What is your opinion of carpooling or ridesharing?
- ▭ A pleasant experience
- ▭ Somewhat pleasant
- ▭ All right
- ▭ Somewhat unpleasant
- ▭ Very unpleasant experience

24. Most people that I am in contact with feel that ridesharing is ...
- ▭ Important
- ▭ Somewhat important
- ▭ No opinion
- ▭ Somewhat unimportant
- ▭ Unimportant

25. How likely are you to share a ride to work in the future?
- ▭ Very likely
- ▭ Somewhat likely
- ▭ Undecided
- ▭ Somewhat unlikely
- ▭ Very unlikely

26. Some employees pay for parking if they drive alone but park for free if they carpool. If it took you 10 minutes extra to be in a carpool, would you drive alone or carpool if the daily parking cost for driving alone was ...

	CARPOOL	DRIVE ALONE
a. $1	▭	▭
b. $2	▭	▭
c. $4	▭	▭
d. $6	▭	▭
e. $8	▭	▭
f. $10	▭	▭

← CONTINUE TO #27

27. Please check how strongly you agree or disagree with each of the following statements:

	STRONGLY DISAGREE → STRONGLY AGREE				
a. The quality of the environment isn't really affected very much by the routine daily activities of individuals	①	②	③	④	⑤
b. We are approaching the limit of people the earth can support	①	②	③	④	⑤
c. The balance of nature is very delicate and easily upset	①	②	③	④	⑤
d. Humans have the right to modify the natural environment to suit their needs	①	②	③	④	⑤
e. Unless we act now to conserve resources, we will soon be faced with a lower quality of life	①	②	③	④	⑤
f. When humans interfere with nature, it often produces disastrous consequences	①	②	③	④	⑤
g. Plants and animals exist primarily to be used by humans	①	②	③	④	⑤
h. The air of our cities would be cleaner if people stopped driving alone to work	①	②	③	④	⑤
i. To maintain a healthy economy we will have to develop a "steady-state" economy where industrial growth is controlled	①	②	③	④	⑤
j. Humans must live in harmony with nature in order to survive	①	②	③	④	⑤
k. The earth is like a spaceship with only limited room and resources	①	②	③	④	⑤
l. There are limits to growth beyond which our industrial society cannot expand	①	②	③	④	⑤
m. Humans need not adapt to the natural environment because they can remake it to suit their needs	①	②	③	④	⑤
n. Mankind was created to rule over the rest of nature	①	②	③	④	⑤
o. Mankind is severely abusing the environment	①	②	③	④	⑤

28. How frequently do you do the following?

	NEVER → ALL THE TIME				
a. Contribute time or money to environmental groups such as the Sierra Club or World Wildlife Fund	①	②	③	④	⑤
b. Recycle material, such as newspaper, glass, aluminum, etc.	①	②	③	④	⑤
c. Read an environmental magazine or newsletter	①	②	③	④	⑤
d. Watch "nature" shows on television	①	②	③	④	⑤
e. Participate in outdoor activities such as camping, hiking and bird-watching	①	②	③	④	⑤
f. Talk to family and friends about environmental issues	①	②	③	④	⑤
g. Use recycled materials	①	②	③	④	⑤

CONTINUE TO NEXT PAGE

29. If a way could be found to cover the costs (such as charging people who commute alone) and your company did the coordination or arrangements, how influential would the following factors be in encouraging you to share rides or use public transportation?

	NOT VERY INFLUENTIAL → VERY INFLUENTIAL				
a. You had a guaranteed ride home for an emergency or if you had to work late	①	②	③	④	⑤
b. Vans were provided for those who would vanpool to and from work	①	②	③	④	⑤
c. A carpool locator would help you find others living and working near you	①	②	③	④	⑤
d. You had free preferential parking for carpools at the building where you work	①	②	③	④	⑤
e. Information on transit routes and schedules	①	②	③	④	⑤
f. A shuttle was provided between transit stops and work	①	②	③	④	⑤
g. A shuttle was provided to shopping centers and offices in the area	①	②	③	④	⑤
h. Dependent or day care facilities were near to your place of employment	①	②	③	④	⑤
i. Work hours were more flexible	①	②	③	④	⑤
j. You could work a compressed work week (10 hours for 4 days)	①	②	③	④	⑤
k. You could work at home sometimes using telecommunications	①	②	③	④	⑤
l. Your cost for public transit was paid	①	②	③	④	⑤
m. There were high occupancy vehicle lanes on the freeways	①	②	③	④	⑤
n. You received rewards or gift points for carpooling of riding public transit	①	②	③	④	⑤
o. Other, specify	①	②	③	④	⑤

30. If your company was faced with a fine or loss of government contracts because too many of its employees drove alone, how justified would the company be in:

	NOT VERY JUSTIFIED → VERY JUSTIFIED				
a. Hiring only employees who live near the work site	①	②	③	④	⑤
b. Requiring employees to move closer to work	①	②	③	④	⑤
c. Moving the company closer to public transit	①	②	③	④	⑤
d. Allowing only ridesharers to park in company lots	①	②	③	④	⑤
e. Reducing the amount of parking	①	②	③	④	⑤
f. Fining those who drive alone	①	②	③	④	⑤
g. Other, specify	①	②	③	④	⑤

31. How frequently does your work group or department <u>currently</u> do the following?

	NEVER → ALL THE TIME				
a. Use team process to make significant work-related decisions	①	②	③	④	⑤
b. Get together outside of the office	①	②	③	④	⑤
c. Discuss issues that affect the department	①	②	③	④	⑤
d. Develop work standards and work tempo as a group	①	②	③	④	⑤
e. Celebrate someone's birthday or significant event	①	②	③	④	⑤
f. Talk about the work different people are doing	①	②	③	④	⑤
g. Be aware of the illness of someone's family member	①	②	③	④	⑤
h. Be influenced by your suggestion for a change	①	②	③	④	⑤
i. Influence the way you do things	①	②	③	④	⑤
j. Function as a team in making decisions	①	②	③	④	⑤

32. How frequently do you think your work group or department <u>should</u> do the following?

a. Use team process to make significant work-related decisions	①	②	③	④	⑤
b. Get together outside of the office	①	②	③	④	⑤
c. Discuss issues that affect the department	①	②	③	④	⑤
d. Develop work standards and work tempo as a group	①	②	③	④	⑤
e. Celebrate someone's birthday or significant event	①	②	③	④	⑤
f. Talk about the work different people are doing	①	②	③	④	⑤
g. Be aware of the illness of someone's family member	①	②	③	④	⑤
h. Be influenced by your suggestion for a change	①	②	③	④	⑤
i. Influence the way you do things	①	②	③	④	⑤
j. Function as a team in making decisions	①	②	③	④	⑤

33. Your **gender:** ☐ Female ☐ Male

34. Your **age:**
☐ Under 25 years
☐ 25-34 years   ☐ 45-54 years
☐ 35-44 years   ☐ 55 years or over

35. Highest amount of **education** you completed?
☐ Less than high school graduate   ☐ College graduate
☐ High school graduate (inc. GED)   ☐ Advanced or professional
☐ Some college   degree

36. What is your **job type?**
☐ Service, maintenance   ☐ Technical
☐ Operative   ☐ Sales/associate
☐ Skilled crafts   ☐ Professional
☐ Secretarial, clerical   ☐ Administrator/manager

37. What is your annual **household income?**
☐ Under $25,000   ☐ $50,000 to $74,999
☐ $25,000 to $49,999   ☐ $75,000 or more

0014001

MAKE NO MARKS IN THIS AREA

Bibliography

Ackerman, Bruce, and William Hassler. 1981. *Clean Coal, Dirty Air.* New Haven: Yale University Press.

Adler, Jonathan. 1994. *Evaluating the Employee Commute Option (ECO): Can ECO Make Economic Sense?* Competitive Enterprise Institute Working Paper, April 7.

Albrecht, Don, Gordon Bultena, Eric Hoiberg, and Peter Nowak. 1982. "The New Environmental Paradigm Scale." *Journal of Environmental Education* 13:39–43.

Allison, Graham T. 1971. *Essence of Decision: Explaining the Cuban Missile Crisis.* Boston: Little, Brown.

Anderson, James E. 1997. *Public Policymaking.* 3rd ed. New York: Houghton-Mifflin.

Anton, Thomas. 1989. *American Federalism and Public Policy: How the System Works.* New York: Random House.

Apter, David E. 1977. *Introduction to Political Analysis.* Cambridge, MA: Winthrop Publishers.

Arbuthnot, J. 1977. "The Role of Attitudinal and Personality Variables in the Prediction of Environmental Behavior and Knowledge." *Environment and Behavior* 9:217–232.

Baltimore Sun. 1992. Letter to the Editor, July 22, 14A.

Barrett, Susan, and Michael Hill. 1984. "Policy, Bargaining and Structure in Implementation Theory." *Policy and Politics* 12:219–240.

Beaton, W. P. 1991. *Transportation Control Measures: Commuting Behavior and the Clean Air Act.* CUPR Policy Report no. 9. Piscataway, NJ: Rutgers State University, Center for Urban Policy Research.

Bloksberg, L. M. 1989. "Intergovernmental Relations: Change and Continuity." *Journal of Aging and Social Policy* 1:11–36.

Boeckelman, K. 1992. "The Influence of States on Federal Policy Adoptions." *Policy Studies Journal* 20:365–375.

Bollens, J. C. and H. J. Schmandt. 1975. *The Metropolis: Its People, Politics, and Economics.* New York: Harper and Row.

Bonham, Gordon Scott. 1996. "Effect of Changing Policy on Employer Commute Reduction Programs." Paper presented to the Urban Affairs Association.

Bonham, Gordon Scott, Vicky Moshier Burnor, and Toni Marzotto. 1994. "Impact on the City of Clean Air Requirements to Change Employee Commuting." Paper presented to Urban Affairs Association, March.

———. 1995. "EPA Clean Air Requirements and Employee Commuting in Baltimore." *Journal of Urban Affairs* 17:165–182.

Bonham, Gordon Scott, and Andrew Z. Farkas. 1995. "Employee Commute Options: Differential Gains and Losses." Proceedings of the Baltimore Symposium of Urban Environmental Justice Research and Education. Baltimore: McKeldin Center at Morgan State University, October: 1–13.

Booth, William. 1997. "In L.A., a Clear Day Is a Dream No Longer." *Washington Post,* December 18, A1.

Breyer, Stephen. 1979. "Analyzing Regulatory Failure: Mismatches, Less Restrictive Alternatives and Reform." *Harvard Law Review* 92:547–609.

Brower, David J., David R. Godschalk, and Douglas R. Porter. 1989 *Understanding Growth Management: Critical Issues and a Research Agenda.* Washington, DC: Urban Land Institute.

Browner, Carol M. 1994. "Partners in Protecting the Public." *The Washington Post,* May 30, A15.

Brownstein, Ronald. 1989. "Testing the Limits." *National Journal,* July 29, 1916–1920.

Brownstone, D., and T. Golob. 1992. "The Effectiveness of Ridesharing Incentives: Discrete Choice Models of Commuting in Southern California." *Regional Science and Urban Economics* 22:5–24.

Bryner, Gary C. 1993. *Blue Skies, Green Politics: The Clean Air Act of 1990.* Washington, DC: Congressional Quarterly Press.

Button, Kenneth, and Werner Rothengatter. 1993. "Global Environmental Degradation: The Role of Transport." In David Banister and Kenneth Button, eds., *Transport, the Environment and Sustainable Development.* London: E and FN SPON.

CAAAC (Clean Air Act Advisory Committee). 1995. "Report of ECO Flexibilities Work Group, Subcommittee on Linking Transportation, Energy and Air Quality," April 26.

Caleb, Solomon. 1994. "Head-on Collision: Cut Auto Commuting? Firms and Employees Gag at Clean-Air Plan." *Wall Street Journal,* September 8, A1.

Cambridge Systematic. 1991. *Effects of Demand Management and Land Use on Traffic Congestion: Literature Review.* Prepared for U.S. Department of Transportation, December.

Campbell, Donald. 1969. "Reforms as Experiments." *American Psychologist* 24, no. 4:409–429.

Carson, Rachel. 1962. *Silent Spring.* Greenwich, CT: Fawcett Crest Books.

Center for Suburban and Regional Studies. 1996a. "Effect of the Employee Commute Options Legislation in the Philadelphia Air Quality Region: Workshop Summary, Delaware." Towson, MD: Towson State University, April.

———. 1996b. "Effect of the Employee Commute Options Legislation in the Philadelphia Air Quality Region: Workshop Summary, New Jersey." Towson, MD: Towson State University, May.

Cervero, Robert. 1986. *Suburban Gridlock.* New Brunswick, NJ: Rutgers University Press.

———. 1989. *America's Suburban Centers: The Land-Use Transportation Link.* Winchester, MA: Unwin Hyman.

Chelimsky, Eleanor. 1985. "Old Patterns and New Directions in Program Evaluation." In Eleanor Chelimsky, ed., *Program Evaluation: Patterns and Directions.* Washington, DC: American Society for Public Administration.

Christian, Sue Ellen, and Jerry Crimmins. 1995. "State to Thumb Nose at Tough Commuting Law." *Chicago Tribune,* March 11, 1.

Church, Thomas W., and Robert T. Nakamura. 1993. *Cleaning Up the Mess: Implementation Strategies in Superfund.* Washington, DC: Brookings Institution.

Cobb, Roger W., and Charles D. Elder. 1984. *Participation in American Politics: The Dynamics of Agenda Building.* 2nd ed. Baltimore: Johns Hopkins University Press.

Cohen, Richard E. 1992. *Washington at Work: Back Rooms and Clean Air.* New York: Macmillan.

COMSIS. 1990. *Cost Effectiveness of Travel Demand Management Measures to Relieve Congestion.* Silver Spring, MD: Prepared for U.S. Department of Transportation, February.

———. 1992. *Development Tools for Phase II Regulation XV Ministerial Plan Review Process: Status Report.* Los Angeles, CA: COMSIS. October 14.

———. 1994. "Effectiveness of Employer Management Program." Air Quality and Mobility Workshop Presentation, Baltimore, MD, May 24.

Congressional Quarterly Almanac. 1989. "Clean-Air Bill Moves in Both Chambers." *Congressional Quarterly Almanac* 45:665–674.

———. 1990. "Stricter Controls Enacted on Smog, Cars, Acid Rain." *Congressional Quarterly Almanac* 46:229–278.

Conlan, Timothy J. 1988. *New Federalism: Intergovernmental Reform from Nixon to Reagan.* Washington, DC: Brookings Institution.

Conlan, Timothy J. and David R. Beam. 1992. "Federal Mandates: The Record of Reform and Future Prospects." *Intergovernmental Perspective* 18, no. 4:7–15.

Cope, John G. 1995. "George Jetson and the Tragedy of the Commons: Applying Behavior Analysis to the Problem of Waste Management." *Environment and Behavior* 27, no. 2:117–121.

Costantini, Edmond, and Kenneth Hanf. 1973. *The Environmental Impulse and Its Competitors.* Davis, CA: Institute of Ecology.

Coughlin, Joseph F. 1994. "The Tragedy of the Commons: Defining Traffic Congestion as a Public Problem." In David A. Rochefort and Roger W. Cobb, *The Politics of Problem Definition: Shaping the Policy Agenda.* Lawrence: University Press of Kansas.

Council of State Governments. 1995. *The Book of the States, 1994–95.* Lexington, KY: Council of State Governments.

Crimmins, Jerry, and Peter Kendall. 1995. "U.S. Car-Pool Law? Never Mind." *Chicago Tribune,* January 21, 1.

Cushman, J. H., Jr. 1990. "Smart Cars and Highways to Help Unsnarl Gridlock." *New York Times,* April 12, A16.

Cusumano, Leanna. 1993. "Analysis of the 1990 Clean Air Act's Employee Commute Options Program: A Trip Down the Right Road." *William and Mary Journal of Environmental Law* 18: 175–218.

Dalton, William O. 1993. "Letter to William Donald Schaefer, Governor of Maryland." December 3.

Davidson, Roger H., and Walter Oleszek. 1990. *Congress and Its Members.* 3rd ed. Washington, DC: Congressional Quarterly Press.

de Leon, Peter. 1983. "Policy Evaluation and Program Termination." *Policy Studies Review* 2, no. 4: 631–648.

DeLeon, Richard E. 1992. "The Urban Antiregime: Progressive Politics in San Francisco." *Urban Affairs Quarterly* 27, no. 4:555–579.

Dodd, Lawrence C., and Bruce I. Oppenheimer. 1985. *Congress Reconsidered.* 4th ed. Washington, DC: Congressional Quarterly Press.

Dodd, Lawrence, and Richard Schott. 1979. *Congress and the Administrative State.* New York: John Wiley.

Dodson, A. L., and K. J. Mueller. 1993. "National Health Care Reform: Wither State Governments?" *Policy Currents* 3:7–26.

Dorgan, Charity Anne, ed. 1995. *Statistical Record of the Environment.* 3rd ed. Detroit, MI: Gale Research.

Downs, Anthony. 1972. "Up and Down with Ecology: The Issue Attention Cycle." *The Public Interest* 28:38–50.

———. 1992. *Stuck in Traffic. Coping with Peak-Hour Traffic Congestion.* Washington, DC: Brookings Institution.

Dunlap, Riley E. 1995. "Public Opinion and Environmental Policy." In James P. Lester, ed. *Environmental Politics and Policy: Theories and Evidence.* Durham, NC: Duke University Press: 63–114.

Dunlap, Riley, and Angela Mertig. 1992. *American Environmentalism.* Philadelphia: Taylor and Francis.

Dunlap, Riley, and Rik Scarce. 1991. "The Polls—Poll Trends: Environmental Problems and Protection." *Public Opinion Quarterly* 55:651–672.

Dunlap, Riley, and Kenneth Van Liere. 1978. "The New Environmental Paradigm." *Journal of Environmental Education* 9:10–19.

———. 1984. "Commitment to the Dominant Social Paradigm and Concern for Environmental Quality." *Social Science Quarterly* 65:1013–1028.

Dunn, William N. 1994. *Public Policy Analysis: An Introduction.* 2nd ed. Englewood Cliffs, NJ: Prentice-Hall.

Dye, Thomas R. 1984. *Understanding Public Policy.* 5th Edition. Englewood Cliffs, NJ: Prentice-Hall.

———. 1990. *American Federalism: Competition Among the States.* Lexington, MA: Lexington Books.

Easton, David. 1965. *A Systems Analysis of Political Life.* New York: John Wiley.

Edelman, Murray Jacob. 1967. *The Symbolic Uses of Politics.* Urbana, IL: University of Illinois Press.

———. 1971. *Politics as Symbolic Action.* Chicago: Markman.

Edwards, George C. III. 1980. *Implementing Public Policy.* Washington, DC: Congressional Quarterly Press.

Edwards, George C. III, and Stephen J. Wayne. 1997. *Presidential Leadership: Politics and Policy Making.* New York: St. Martins.

Elmore, Richard. 1979. "Backward Mapping: Using Implementation Analysis to Structure Political Decisions." *Political Science Quarterly* 94:601–616.

———. 1982. "Backward Mapping: Implementation Research and Policy Decisions." In Walter Williams et al. *Studying Implementation: Methodological and Administrative Issues.* Chatham, NJ: Chatham House Publishers: 18–35.

———. 1987. "Instruments and Strategy in Public Policy." *Policy Studies Review* 7, no. 1:174–186.

Environmental Law Institute. 1996. *Fifth Anniversary of the 1990 Amendments to the Clean Air Act: Symposium Proceedings.* Washington, DC: Environmental Law Institute.

Ernst and Young. 1992. "Rule 1501 Cost Survey." *South Coast Air Quality Management District* (August).

Eulau, Heinz. 1963. *The Behavioral Persuasion in Politics.* New York: Random House.

Eulau, Heinz, and Kenneth Prewitt. 1973. *Labyrinths of Democracy: Adaptations, Linkages, Representation, and Policies of Urban Politics.* Irvington, IN: Bobbs-Merrill.

Ewing, R. 1993. "TDM, Growth Management and the Other Four Out of Five Trips." *Transportation Quarterly* (July): 343–366.

Farkas, Z. Andrew. 1996. "The Equity and Cost Effectiveness of Employee Commute Options Programs." Paper presented at the Transportation Research Board 75th Annual Meeting, Washington, DC, January.

Federal Highway Administration. 1992. *Our Nation's Highways.* Washington, DC: U.S. Department of Transportation.

———. 1994. *Journey to Work Trends in the United States and Its Major Metropolitan Areas, 1960–1990.* Washington, DC: U.S. Department of Transportation.

Fox, Stephen. 1981. *John Muir and His Legacy: The American Conservation Movement.* Boston: Little, Brown.

Friedlaender, Anne, ed. 1978. *Approaches to Controlling Air Pollution.* Cambridge, MA: MIT Press.

Friedrich, Carl J. 1963. *Man and His Government.* New York: McGraw-Hill.

Gage, Robert W., and Bruce D. McDowell. 1995. "ISTEA and the Role of MPOs in the New Transportation Environment: A Midterm Assessment." *Publius: The Journal of Federalism* 25, no. 3: 133–154.

Garreau, J. 1991. *Edge City.* New York: Doubleday.

Gavin, William. 1992. *Letter to Susan Wierman, Acting Director, Maryland Department of Environment.* October 13.

Geller, F. Scott. 1991. "If Only More Would Actively Care." *Journal of Applied Behavior Analysis* 24, no. 4: 607–612.

———. 1995. "Actively Caring for the Environment: An Integration of Behaviorism and Humanism." *Environment and Behavior* 27 (March):184–195.

Geller, Jack M., and Paul Lasley. 1985. "The New Environmental Paradigm Scale: A Reexamination." *Journal of Environmental Education* 17, no. 1:9–12.

Gerson, Larry N. 1997. *Public Policy Making: Process and Principles.* Armonk, NY: M. E. Sharpe.

Giuliano, G. 1992a. "Transportation Demand Management and Urban Traffic Congestion: Promise or Panacea?" *Journal of the American Planning Association* 58, no. 3:327–335.

———. 1992b. "An Assessment of the Political Acceptability of Congestion Pricing." *Transportation* 19:335–358.

Giuliano, Genevieve, and Martin Wachs. 1992. "Responding to Congestion and Traffic Growth: Transportation Demand Management." In J. Stein, ed., *Growth Management and Sustainable Development.* Newbury Park, CA: Sage Publications.

Giuliano, G., K. Hwang, and Martin Wachs. 1993. "Employee Trip Reduction in Southern California: First Year Results." *Transportation Research—A,* 27A, no. 2:125–137.

Goggins, Malcolm L. 1987. *Policy Design and the Politics of Implementation.* Knoxville: University of Tennessee Press.

Gormley, William. 1989. *Taming the Bureaucracy.* Princeton, NJ: Princeton University Press.

Graham, Allison. 1971. *Essence of Decision.* Boston: Little, Brown.

Green, Kenneth. 1994. "Looking Beyond ECO: Alternatives to Employer-Based Trip Reduction." Reason Foundation. Policy Study no. 185.

———. 1995. "Defending Automobility: A Critical Examination of the Environmental and Social Costs of Auto Use." Reason Foundation. Policy Study no. 198.

Greer, Charles R., and H. Kirk Downey. 1982. "Industrial Compliance with Social

Legislation: Investigations of Decision Rationales." *Academy of Management Review* 7, no. 3: 488–498.

Grube, Craig A. 1993. "Letter to Merrylin Zaw-Mon, Director, Air and Radiation Management Administration, Maryland Department of Environment." June 23.

Hardin, G. 1968. "Tragedy of the Commons." *Science.* (December):1243–1248.

Hargrove, Erwin C. 1976. *The Missing Link: The Study of the Implementation of Social Policy.* Washington, DC: Urban Institute.

Hawthorn, Gary. 1988. *The Role for Transportation Control Measures in the Post '87 Era.* Washington, DC: EPA.

———. 1996a. "State Responses to CAA ECO Amendment." *LINKS—News and Analysis of Transportation and Air Quality Issues* (March).

———. 1996b. Telephone interview. Editor, *LINKS—News and Analysis of Transportation and Air Quality Issues.* March 15.

———. 1997. Telephone interview. November 19.

Hayes, Michael T. 1992. *Incrementalism and Public Policy.* New York: Longman.

Hays, Samuel P. 1987. *Beauty, Health, and Permanence: Environmental Politics in the United States, 1955–1985.* Cambridge: Cambridge University Press.

Heclo, Hugh. 1978. "Issue Networks and the Executive Establishment." In Anthony King, ed., *The New American Political System.* Washington, DC: American Enterprise Institute.

Hjern, Benny. 1982. "Implementation Research—the Link Gone Missing." *Journal of Public Policy* 2, no. 3: 301–308.

Hjern, Benny, and Chris Hull. 1982. "Implementation Research as Empirical Constitutionalism." *European Journal of Political Research* 10:105–116.

Hjern, Benny, and David Porter. 1981. "Implementation Structures." *Organization Studies* 2:211–227.

Hogwood, Brian W., and B. Guy Peters. 1985. *The Pathology of Public Policy.* Oxford: Oxford University Press.

Honadale, Beth. 1981. "A Capacity Building Framework: A Search for Concept and Purpose." *Public Administration Review* 41: 577–589.

Ibata, David. 1994a. "Commute Mandates Under Fire." *Chicago Tribune,* January 12, 2C6.

———. 1994b. "Foes of Clean-Air Act in Overdrive: Business Groups Fear High Costs Will Outweigh Benefits." *Chicago Tribune,* February 9, 6.

———. 1994c. "Local Firms Buck U.S. Law as Clean Air Deadline Nears." *Chicago Tribune,* January 7, 1.

———. 1994d. "Remove Mandate on Car Pools from Clean Air Act, Edgar Says." *Chicago Tribune,* March 6, 1.

———. 1994e. "State Business Groups Urge Lawmakers to Fight Commute Rules in Clean-Air Law." *Chicago Tribune,* January 12, 6.

Ingram, Helen, and Anne Schneider. 1988. "Improving Implementation Through Framing Smarter Statutes." *International Public Policy* 10, no. 1:67–88.

Jaffe, Mark. 1993. "Getting Tough on Pollution Rules: EPA Has Denied Bids for Waivers of Clean-Air Rules." *Philadelphia Inquirer,* December 18, B1.

———. 1995. "At DER, Ridge Plans to Turn over a New Leaf." *Philadelphia Inquirer,* April 4, B1.

Jenkins-Smith, Hank, and Paul Sabatier. 1994. "Evaluating the Advocacy Coalition Framework." *Journal of Public Policy* 14:175–203.

Jennings, Deborah, E. 1994. "Employee Commute Options Come Back to Life." *Maryland Environmental Law Letter* 2:1–3.

Jensen, P. 1993. "More Commuters Now Working in Suburbs." *Baltimore Sun,* (January 6), 1B.

Johnson, David W. et al. 1981. "The Effects of Cooperative, Competitive and Individualistic Goal Structures on Achievement: A Meta-Analysis." *Psychological Bulletin* 89, no. 1:47–62.

Johnson, E. 1992. Project Report: The Future of the Automobile in the Urban Environment. *Bulletin of the American Academy of Arts and Sciences,* 45:7–22.

Jones, Charles. 1975. *Clean Air.* Pittsburgh: University of Pittsburgh Press.

———. 1977. *An Introduction to the Study of Public Policy.* 3rd ed. Belmont, CA: Wadsworth.

Kamieniecki, Sheldon, and Michael R. Ferrall. 1991. "Intergovernmental Relations and Clean-Air Policy in Southern California." *Publius: The Journal of Federalism* 21: 143–154.

Kangun, Norman, and R. Charles Moyer. 1976. "The Failings of Regulation." *Michigan State University Business Topics* 24, no. 2: 5–14.

Kaplan, Marshal. 1995. "Urban Policy: An Uneven Past, an Uncertain Future." *Urban Affairs Quarterly* 30, no. 5:662–680.

Kelman, Steven. 1996. *American Democracy and the Public Good.* Fort Worth, TX: Harcourt Brace College Publishers.

Kenyon, Daphne, and John Kincaid. 1991. *Competition Among State and Local Governments: Efficiency and Equity in American Federalism.* Washington, DC: Urban Institute.

Kerwin, Cornelius M. 1994. *Rulemaking: How Government Agencies Write Law and Make Policy.* Washington, DC: Congressional Quarterly Press.

Kincaid, John. 1990. "From Cooperative to Coercive Federalism." *The Annals of the American Academy of Political and Social Sciences* 509:148–152.

Kingdon, John W. 1984. *Agendas, Alternatives and Public Policies.* Boston: Little, Brown.

Kingsley, Game. 1979. "Controlling Air Pollution: Why Some States Try Harder." *Policy Studies Journal* 7: 728–738.

Kirkwood, John L. 1994. "A Sound Program for Cleaning the Air." *Chicago Tribune,* March 22, 20.

Kraft, Michael E. 1996. *Environmental Policy and Politics.* New York: Harper Collins.

Kraft, Michael E., and Norman J. Vig. 1994. "Environmental Policy from the 1970s to the 1990s: An Overview." In Norman J. Vig and Michael E. Kraft, eds., *Environmental Policy in the 1990s: Towards a New Agenda.* Washington DC: Congressional Quarterly Press: 1–29.

Kuzmyzk, J. Richard. 1992. *Employer Managed Transportation: Twenty Years of Managing Travel Demand at 3M.* Washington, DC: Surface Transportation Policy Project, August.

Lane, R. 1993. "The Commuter Police." *Forbes,* December 29, 239–240.

Langfitt, F.C. 1993. Pollution Restrictions Criticized: Schaefer Decries U.S. Requirements. *Baltimore Sun,* October 11, 1B.

Ledebur, Laurence C., and William R. Barnes. 1992. *Metropolitan Disparities and Economic Growth: City Distress and the Need for a Federal Local Growth Package.* Washington, DC: National League of Cities.

Lester, James P., and Ann O'M. Bowman. 1989. "Implementing Environmental Policy in a Federal System: A Test of the Sabatier-Mazmanian Model." *Polity* 21:731–53.

Lewin, Kurt. 1951. *Field Theory in Social Science.* New York: Harper and Row.

Lieberman, Joseph I., Frank R. Lautenberg, and Harris Wofford. 1994. "Letter to Ms. Carol M. Browner, Administrator, Environmental Protection Agency." May 31.

Lindblom, Charles. 1959. "The Science of Muddling Through." *Public Administration Review* 19:79–88.

Lippmann, Walter. 1938. *Public Opinion*. New York: Macmillan.

Lipsky, Michael. 1980. *Street Level Bureaucracy: Dilemmas of the Individual in Public Service*. New York: Russell Sage.

Lis, James, and Kenneth Chilton. 1994. "Using the Wrong Measures for Smog." *Regulation* 1: 51–59.

Lowi, Theodore J., and Benjamin Ginsberg. 1996. *American Government*. New York: W. W. Norton.

Manzullo, Donald. 1997. Personal interview. Member of Congress from Illinois. May 20.

Maryland, Department of the Environment (MDE). 1993. "Letter from Merrylin Zaw-Mon, Director, Air and Radiation Management Division to Thomas J. Maslany, Director, Air, Radiation and Toxics Division, EPA." September 30.

———. 1995. *Report on the Employee Commute Options Program*. Baltimore, MD: Maryland Department of Environment. September 15.

Maryland Department of the Environment, Air and Radiation Management Administration (MDEARMA). 1993. *Transcript of Public Hearing: In the Matter of Air-Quality Regulations Found in COMAR 26.11.25, Employee Commute Options*. Baltimore, MD: Maryland Department of Environment.

———. 1994. *Employee Commute Options* 21 (June 10).

Marzotto, Toni. 1996. "Federalism and the Implementation of the Employee Commute Options Program: From Mandates to Options." Paper presented to 1996 Annual Meeting of Urban Affairs Association, New York, March.

———. 1998. "Congestion Pricing: An Idea Whose Time Has Come?" Paper presented to 1998 Annual Meeting of the American Political Science Association, Massachusetts, September.

Marzotto, Toni, Vicky Moshier Burnor, and Gordon Scott Bonham. 1995. "The Changing Environment of Regulatory Compliance: A New Look at the Clean Air Act." Paper presented to 1995 Annual Meeting of the American Society for Public Administration, San Antonio, Texas, July.

May, Peter J. 1995. "Can Cooperation Be Mandated? Implementing Intergovernmental Environmental Management in New South Wales and New Zealand." *Publius: The Journal of Federalism* 25: 89–102.

Mazmanian, Daniel, and Paul Sabatier. 1980. "A Multivariate Model of Public Policy-Making." *American Journal of Political Science* 24: 439–468.

———. 1983. *Implementation and Public Policy*. Glenview, IL: Scott, Foresman.

———. 1989. *Implementation and Public Policy*. Lanham, MD: University Press of America.(Reprint of 1983 book published by Scott Foresman and Co.)

Mazmanian, Daniel, and Paul Sabatier, eds. 1981. *Effective Policy Implementation*. Lexington, MA: DC Heath.

McKelvey, Laura. 1997. Telephone interview. Transportation coordinator, South Coast Air Quality Management District. November 18.

Melnick, R. S. 1992. "Pollution Deadlines and the Coalition for Failure." In Michael Greve and Fred Smith, Jr., eds. *Environmental Politics: Public Costs, Private Rewards*. New York: Praeger.

Meszler, Donald A. 1993. "Letter to Merrylin Zaw-Mon, Director, Maryland Department of the Environment." October 12.

Milbrath, Lester W. 1984. *Environmentalists: Vanguard for a New Society*. Albany: State University of New York Press.

Morris, John A. 1992. "Md Wants 'Pooling' to Save Air: Employers Soon Held Responsible." *Baltimore Sun*, August 19, 2.

Mullaney, Timothy J. 1992. "Driving to Work Could Mean Loss of a Job." *Baltimore Sun,* July 18, 12C.

Nakamura, Robert. 1987. "The Textbook Policy Process and Implementation Research." *Policy Studies Review* 1:142–154.

Nakamura, Robert, and Frank Smallwood. 1980. *The Politics of Policy Implementation.* New York: St. Martin's.

Nathan, Richard P., Fred C. Doolittle, and associates. 1987. *Reagan and the States.* Princeton, NJ: Princeton University Press.

National Performance Review. 1995. *Common Sense Government: Works Better and Costs Less.* Washington, DC: Government Printing Office.

New Jersey Register. 1994. 26 NJR 756 (February 7).

Newhouse, Nancy. 1990. "Implications of Attitude and Behavior Research for Environmental Conservation." *Journal of Environmental Education* 22, no. 1:26–30.

Nichols, Mary D. 1996. "Consensus in Drafting, Smooth Implementation." *The Environmental Forum* 13:41.

Noe, Francis P., and Rob Snow. 1990. "The New Environmental Paradigm and Further Scale Analysis." *Journal of Environmental Education* 21, no. 4:20–26.

O'Connor, Karen. 1996. *No Neutral Ground? Abortion Politics in an Age of Absolutes.* Boulder, CO: Westview.

O'Connor, Karen, and Larry Sabato. 1995. *American Government: Roots and Reform.* 2nd ed. Boston, MA: Allyn and Bacon.

Orski, C. Kenneth. 1990. "Can Management of Transportation Demand Help Solve Our Growing Traffic Congestion and Air Pollution Problems?" *Transportation Quarterly* 44:483–498.

———. 1993. "Employee Trip Reduction Programs: An Evaluation." *Transportation Quarterly* 47:327–341.

———. 1995. Telephone interview. President, Urban Mobility Corporation, Washington, DC. November 9.

Osborne, David E. 1988. *Laboratories of Democracy: A New Breed of Governor Creates Models for National Growth.* Boston, MA: Harvard Business School Press.

Palumbo, Dennis. 1988. *Public Policy in America.* New York: Harcourt, Brace, Jovanovich.

Patton, Michael Quinn. 1997. *Utilization-Focused Evaluation.* 3rd ed. Thousand Oaks, CA: Sage Publications.

Peirce, Neil. 1993. *Citistates: How Urban America Can Prosper in a Competitive World.* Washington, DC: Seven Locks Press.

Pennsylvania Department of Environmental Resources. 1995. *Employer Trip Reduction Program Policy.* February 27.

Peters, Guy B. 1996. *American Public Policy: Promise and Performance.* 4th ed. Chatham, NJ: Chatham House.

Peters, Guy, and Brian Hogwood. 1985. "In Search of the Issue Attention Cycle." *Journal of Politics* 47:238–253.

Peterson, Paul E. 1995. *The Price of Federalism.* Washington, DC: Brookings Institution.

Petracca, Mark P. 1992. *The Politics of Interests.* Boulder, CO: Westview.

Pickrell, D. 1985. "Rising Deficits and the Uses of Transit Subsidies in the United States." *Journal of Urban Economics* 19:281–298.

Pisarski, Alan. 1987. *Commuting in America: A National Report on Commuting Patterns and Trends.* Westport, CT: Eno Transportation Foundation.

————. 1996. *Commuting in America II: The Second Report on Commuting Patterns and Trends*. Lansdowne, VA: Eno Transportation Foundation.

Portney, Paul R. 1990. "Air Pollution Policy." In Paul R. Portney, ed., *Public Policies for Environmental Protection*. Washington, DC: Resources for the Future.

Post, J. E. 1978. "The Corporation in the Public Policy Process—A View Toward the 1980s." *Sloan Management Review* 21, no. 1: 45–52.

Pressman, Jeffrey, and Aaron Wildavsky. 1973. *Implementation*. Berkeley, CA: University of California Press.

Public Law 84–159. 1955. *Air Pollution Control Act*. 84th Congr., 1st sess., July 14.

Public Law 88–206. 1963. *Clean Air Act*. 88th Congr., 1st sess., December 17.

Public Law 89–272. 1965. *Motor Vehicle Air Pollution Control Act*. 89th Congr., 1st sess., October 20.

Public Law 90–148. 1967. *Air Quality Act*. 90th Congr., 1st sess., November 21.

Public Law 91–604. 1970. *Clean Air Act*. 91st Congr., 2nd sess., December 31.

Public Law 95–95. 1977. *Clean Air Act Amendments*. 95th Congr., 1st sess., August 7.

Public Law 101–549. 1990. *Clean Air Act Amendments*. 101st Congr., 2nd sess., November 15.

Public Law 103–172. 1993. *Federal Employees Clean Air Act Incentives Act*. 103rd Congr., 1st sess., December 2.

Rechtshaffen, Joyce. 1994. Telephone interview. Legislative Assistant, Senator Joseph Lieberman. February 15.

Renner, Michael. 1989. "Cars and Pollution: Rethinking Transportation." *Current* (June): 32–40.

Rochefort, David A., and Roger W. Cobb. 1994. *The Politics of Problem Definition: Shaping the Policy Agenda*. Lawrence, KS: University Press of Kansas.

Rodgers, Harrell, and Charles Bullock. 1976. *Coercion to Compliance*. Lexington, MA: DC Heath.

Roosevelt University. 1992. *Putting Federalism to Work in America: Tracking the Problem of Unfunded Mandates and Burdensome Regulations*. Chicago, IL: Roosevelt University.

Rose, Richard, ed. 1969. *Policy Making in Great Britain*. London: Macmillan.

Rosenbaum, Walter A. 1995. *Environmental Politics and Policy*. 3rd ed. Washington, DC: Congressional Quarterly Press.

Rossi, Peter H., and Howard E. Freeman. 1989. *Evaluation: A Systematic Approach*. 4th ed. Newbury Park, CA: Sage Publications.

Rusk, David. 1993. *Cities Without Suburbs*. Washington, DC: Woodrow Wilson Center Press.

Ruth, Connie. 1997. Telephone interviews. Environmental Protection Specialist. U.S. Environmental Protection Agency, Office of Mobile Sources. March 4, May 12, and November 18.

Sabatier, Paul A. 1986. "Top-Down and Bottom-Up Models of Policy Implementation: A Critical Analysis and Suggested Synthesis." *Journal of Public Policy* 6:21–48.

————. 1987. "Knowledge, Policy-Oriented Learning, and Policy Change." *Knowledge: Creation, Decision, Utilization* 8:649–692.

————. 1988. "An Advocacy Coalition Framework of Policy Change and the Role of Policy-Oriented Learning Therein." *Policy Sciences* 21:129–168.

————. 1993. "Policy Change over a Decade or More." In Paul A. Sabatier and Hank Jenkins-Smith. *Policy Change and Learning: An Advocacy Coalition Approach*. Boulder, CO: Westview, 13–40.

Sabatier, Paul A., and Hank Jenkins-Smith. 1993. "The Advocacy Coalition Framework: Assessment, Revisions, and Implications for Scholars and Practitioners." In Sabatier, and Jenkins-Smith, *Policy Change and Learning: An Advocacy Coalition Approach.* Boulder, CO: Westview, 211–236.

Sabatier, Paul A., and Daniel A. Mazmanian. 1979. "The Conditions of Effective Implementation: A Guide to Accomplishing Policy Objectives." *Policy Analysis* 5: 481–504.

Sachs, Adam. 1992. "Shared Commutes Are the Way to Go: More Find Carpools Ease Traffic Pollution." *Baltimore Sun,* May 10, 2.

Samdahl, D. M., and R. Robertson. 1989. "Social Determinants of Environmental Concern." *Environment and Behavior* 21, no. 1:57–81.

Samuelson, C. D., and M. Biek. 1991. "Attitudes Toward Energy Conservation: A Confirmatory Factor Analysis." *Journal of Applied Social Psychology* 21, no. 7:550.

Schlozman, Kay Lehman, and John T. Tierney. 1986. *Organized Interests and American Democracy.* New York: Harper Press.

Scholtz, John T. 1991. "Cooperative Regulatory Enforcement and the Politics of Administrative Effectiveness." *American Political Science Review* 85, no. 1:115–136.

Shoup, Donald, and Richard Willson. 1992. "Employer-Paid Parking: The Problem and Proposed Solutions." *Transportation Quarterly* 46:169–192.

Shoup, Donald, and Don Pickrell. 1980. *Free Parking as a Transportation Problem.* Washington, DC: U.S. Department of Transportation.

Shrouds, James M. 1995. "Challenges and Opportunities for Transportation: Implementation of the Clean Air Act Amendments of 1990 and the Intermodal Surface Transportation Efficiency Act of 1991." *Transportation* 22:193–215.

Sivek, David J., and Harold Hungerford. 1990. "Predictors of Responsible Behavior in Members of Three Wisconsin Conservation Organizations." *Journal of Environmental Education* 21, no. 2:35.

Smith, Zachary. 1992. *The Environmental Policy Paradox.* Englewood Cliffs, NJ: Prentice Hall.

Sport Truck. 1989. Editorial.

Stark, Rodney. 1992. *Sociology.* 4th ed. Belmont, CA: Wadsworth Publishing.

Starling, Grover. 1998. *Managing the Public Sector.* 5th ed. Fort Worth, TX: Harcourt Brace.

Stewart, J. 1994. "Current Problems in Performing Accurate Cost-Benefit Evaluations of Employer-Based Vehicle Trip Reduction Programs." Paper presented at 73rd Annual Meeting of the Transportation Research Board, Washington, DC, January 9–13.

Stranahan, Susan Q. 1992. "Commuting to Work Alone? Think Again." *Philadelphia Inquirer,* October 25, A1.

———. 1993. "Area Firms Will Fight EPA Commuter Plan." *Philadelphia Inquirer,* June 12, A1.

———. 1994. "Commuting Plan Starts, Answers Lag." *Philadelphia Inquirer,* Feburary 2, A1.

Struzzi, Diane. 1992. "A Welcome Concept for Commuting Soon Will Become the Law." *Philadelphia Inquirer,* July 10, B6.

Surber, Monica, Donald Shoup, and Martin Wachs. 1984. "Effects of Ending Employer-Paid Parking for Solo Drivers." *Transportation Research Record* 957:67–71.

Switzer, Jacqueline Vaughn. 1997. *Green Backlash: The History and Politics of Environmental Opposition in the U.S.* Boulder, CO: Lynne Rienner Publishers.

Tjosvold, D. 1984. "Cooperation Theory and Organizations." *Human Relations* 37, no. 9:743–767.

Truman, David. 1951. *The Governmental Process*. New York: Alfred Knopf.

Urban Mobility Corporation. 1994. *Innovation Briefs* 5:1–3.

———. 1995. "The Lessons of California's Experience." *Innovation Briefs* 6:1–2.

U.S. Congress. House Committee on Energy and Commerce, Subcommittee on Health and Environment. 1989a. *Clean Air Act Standards*. 101st Cong., 1st sess., February 28.

———. 1989b. *Clean Air Act Reauthorization* (Part 1). 101st Cong., 1st sess., September 7.

———. 1989c. *Clean Air Act Reauthorization* (Part 2). 101st Cong., 1st sess., September 12; October 4 and 11.

———. 1989d. *Clean Air Act Reauthorization* (Part 3). 101st Cong., 1st sess., October 18–19.

———. 1989e. *Clean Air Act Amendments* (Part 1). 101st Cong., 1st sess., May 23.

———. 1989f. *Clean Air Act Amendments* (Part 2). 101st Cong., 1st sess., May 24.

———. 1989g. *Clean Air Act Amendments* (Part 3). 101st Cong., 1st sess., June 22; July 24.

U.S. Congress. House Committee on Public Works and Transportation, Subcommittee on Investigation and Oversight. 1989h. *The Impact of Increased Air Quality Regulation on Federal Highway and Transit Programs, and on Fuel Tax Collection*. 101st Cong., 1st sess., November 9.

U.S. Congress. House Committee on Public Works and Transportation. 1990a. *Provisions of H.R. 3030 The Clean Air Act Amendments of 1989 That Fall Within the Jurisdiction of the Committee on Public Works and Transportation*. 101st Cong., 2nd sess., April 19.

U.S. Congress. House Conference Report to Accompany S. 1630. 1990b. *Clean Air Act Amendments of 1990*. 101st Cong., 2nd sess., October 26.

U.S. Congress. House. 1994. "Introductory Remarks of Representative Don Manzullo in Support of H.R. 4589, a Bill to Amend the Clean Air Act." 103rd Cong., 2nd sess. *Congressional Record* 140 (June 16), pt. 1.

U.S. Congress. House Committee on Commerce, Subcommittee on Oversight and Investigations. 1995a. *Implementation and Enforcement of Clean Air Act Amendments of 1990*. 104th Cong., 1st sess., February 9, March 16.

U.S. Congress. House. 1995b. "Speaking for Passage of H.R. 325, a Bill to Amend the Clean Air Act." 104th Cong., 1st sess. *Congressional Record* 141 (March 13), pt. 2.

U.S. Congress. House. 1996. *A Bill to Amend the Clear Air Act*. H.R. 3519. 104th Cong. 2nd sess. *Congressional Record* 142 (May 23), pt. 1.

U.S. Congress, OTA (Office of Technology Assessment). 1989. *Catching Our Breath: Next Steps for Reducing Urban Ozone*. OTA-O–412. Washington, DC: U.S. GPO.

U.S. Congress. Senate Committee on Environment and Public Works, Subcommittee on Environmental Protection. 1989a. *Clean Air Act Amendments of 1989* (Part 1). 101st Cong., 1st sess., September 21.

———. 1989b. *Clean Air Act Amendments of 1989* (Part 2). 101st Cong., 1st sess., September 26.

———. 1989c. *Clean Air Act Amendments of 1989* (Part 3). 101st Cong., 1st sess., September 27.

———. 1989d. *Clean Air Act Amendments of 1989* (Part 4). 101st Cong., 1st sess., September 28.

U.S. Congress. Senate Committee on Environment and Public Works. 1989e. *Clean Air Act Amendments of 1989.* 101st Cong., 1st sess., December 20.

U.S. Congress. Senate. 1995. "Senator Chafee of Rhode Island, Speaking for Passage of H.R. 325, a Bill to Amend the Clean Air Act." 104th Cong., 1st sess. *Congressional Record* 141 (March 13), pt. 2.

U.S. Department of Commerce (USDC). 1994. *Statistical Abstracts of the U.S. 1994.* Washington, DC: Government Printing Office.

———. 1995. *Statistical Abstracts of the United States, 1995.* Washington, DC: Government Printing Office.

———. 1998. *Statistical Abstracts of the United States, 1998.* Washington, DC: Government Printing Office.

USEPA (U.S. Environmental Protection Agency). 1973. *The Clean Air Act and Transportation Controls: An EPA White Paper.* Washington, DC: USEPA.

———. 1992. *Employee Commute Options Guidance.* Washington, DC: USEPA, Office of Air and Radiation (ANR–443), December 20.

———. 1993a. *Employee Commute Options Programs: Status Sheet.* Washington, DC: USEPA, Office of Air and Radiation, December.

———. 1993b. "Letter from Thomas J. Maslany, Director Air, Radiation & Toxics Division, Region III, to Ms. Merrylin Zaw-Mon, Director, Maryland Department of the Environment." October 18.

———. 1993c. "Letter from James Hemby, Chief, Office of Program Integration, Region III, to Ms. Susan S.G. Wierman, Deputy Director, Maryland Department of the Environment." June 7.

———. 1993d. "Letter from Thomas J. Maslany, Director Air, Radiation & Toxics Division, Region III, to Ralph Raab, Delaware Transportation Authority." May 18.

———. 1993e. "Letter from Valdas V. Adamkus, Regional Administrator, Region 5, to Kirk Brown, Secretary, Illinois Department of Transportation." September 3.

———. 1993f. "Memorandum from Conrad Simon, Director Air and Waste Management Division, Region II, to Richard Wilson, Director Office of Mobile Sources regarding Employee Commute Options Program Issues." December 6.

———. 1994. "Letter from Carol M. Browner, Administrator, to the Honorable Frank R. Lautenberg, United States Senate." June 19.

———. 1995a. "Memorandum from Margo T. Oge, Director, Office of Mobile Sources, to Air Division Directors, Regions 1–10, Regarding Additional Options for Flexibility in the Employee Commute Options Program." August 18.

———. 1995b. "Memorandum from Mary D. Nichols, Assistant Administrator for Air and Radiation, to Air Division Directors, Regions 1–10, Regarding EPA Response to Clean Air Act Advisory Committee ECO Recommendations." (June 29).

———. 1996. Air and Radiation. Office of Mobile Sources. *Episodic Emission Control Programs.* December.

———. 1997. Air and Radiation. Office of Mobile Sources. *Episodic Emission Control Programs.* October 23.

USGAO (General Accounting Office). 1988. *Air Pollution: Ozone Attainment Requires Long-Term Solutions to Solve Complex Problems.* Washington, DC: GAO.

Van Meter, Donald S., and Carl E. Van Horn. 1975. "The Policy Implementation Process: A Conceptual Framework." *Administration and Society* 6, no. 4.

Vig, Norman J. 1994. "Presidential Leadership and the Environment: From Reagan and Bush to Clinton." In Norman J. Vig and Michael E. Kraft, eds. *Environmental Policy in the 1990s: Towards a New Agenda.* Washington, DC: Congressional Quarterly Press.

Wachs, Martin, and Genevieve Giuliano. 1992. "Employee Transportation Coordinators: A New Profession in Southern California." *Transportation Quarterly* 46:411–427.

Walker, David B. 1995. *The Rebirth of Federalism: Slouching Toward Washington.* Chatham, NJ: Chatham House.

Wall Street Journal. 1994a. "Call Off the Dogs." Editorial, November 16, A28.

———. 1994b. "Hitting the Carpool Brakes." Editorial, January 12, A10.

Webber, David. 1983. "Obstacles to the Utilization of Systematic Policy Analysis." *Knowledge* 4:534–560.

———. 1986. "Explaining Policymakers' Use of Policy Information." *Knowledge* 7:249–290.

Webber, Melvin M. 1992. "The Joys of Automobility." In Martin Wachs and Margaret Crawford, eds., *The Car and the City: The Automobile, the Built Environment, and Daily Urban Life.* Ann Arbor: University of Michigan Press, 254–273.

Weiss, Carol. 1972. *Evaluation Research: Methods for Assessing Program Effectiveness.* Englewood Cliffs, NJ: Prentice-Hall.

Wheeler, Timothy B. 1994a. "Car Pool Rules May Be Eased." *Baltimore Sun,* April 15, B1.

———. 1994b. "Exemptions Sought for Commuting Rules." *Baltimore Sun,* January 12, 5B.

———. 1995. "Glendening Suspends State's Participation in Troubled Program to Cut Smog." *Baltimore Sun,* May 31, 4B.

Wholey, Joseph S. 1970. *Federal Evaluation Policy.* Washington, DC: Urban Institute.

———. 1983. *Evaluation and Effective Public Management.* Boston, MA: Little, Brown.

Wholey, Joseph S., Harry P. Hatry, and Kathryn E. Newcomer. 1994. *Handbook of Practical Program Evaluation.* San Francisco, CA: Jossey-Bass Publishers.

Willson, Richard, and Donald Shoup. 1990. "Parking Subsidies and Travel Choices: Assessing the Evidence." *Transportation* 17:141–158.

Wright, C. C. 1992. *Fast Wheels, Slow Traffic: Urban Transportation Choices.* Philadelphia, PA: Temple University Press.

Yates, Brock. 1988. "Would You Draw to This Pair?" *Car and Driver* 35, no. 2 (November).

Yin, Robert K. 1985. *Case Study Analysis.* Beverly Hills, CA: Sage Publications.

Yuhnke, Robert E. 1991. "The Amendments to Reform Transportation Planning in the Clean Air Act Amendments of 1990." *Tulane Environmental Law Journal* 5:239–253.

Index

About the Book

How is U.S. public policy made? This comprehensive survey, designed to help students and scholars understand the complexity of policymaking, traces the employee commute option (ECO) step by step from initial idea through enactment and implementation to evaluation and reformulation.

The authors integrate two dominant theories in the policy analysis literature—the policy cycle model and the advocacy coalition framework—using them to examine transportation options mandated by the Clean Air Act Amendments of 1990. They also analyze the social and economic impact of environmental policy on private companies and their employees when states attempt to clean the air and reduce congestion by legislating against the love affair between U.S. citizens and their cars. Drawing on original documents and rich data on ECO, this study is the perfect introduction to the policymaking process.

Toni Marzotto is professor of political science at Towson University. **Vicky Moshier Burnor** is research associate for the Regional Economics Studies Institute at Towson University. **Gordon Scott Bonham** is assistant director of the Regional Economics Studies Institute at Towson University.